Horizontal Gene Transfer in Microorganisms

Edited by

M. Pilar Francino

Center for Public Health Research
Valencia
Spain

Caister Academic Press

Caister Academic Press
Norfolk, UK

www.caister.com

British Library Cataloguing-in-Publication Data
A catalogue record for this book is available from the British Library

ISBN: 978-1-908230-10-2

Cover design adapted from Figure 7.4.

Printed and bound in Great Britain

Contents

Contributors

Rustam I. Aminov
Rowett Institute of Nutrition and Health
University of Aberdeen
Aberdeen
UK

r.aminov@abdn.ac.uk

Miriam Barlow
School of Natural Sciences
University of California
Merced, CA
USA

miriam.barlow@gmail.com

Eva C. Berglund
Department of Medical Sciences
Uppsala University
Uppsala
Sweden

eva.berglund@medsci.uu.se

Luís M. Carvalho
Centro de Biologia Ambiental (CBA)
Faculdade de Ciências da Universidade de Lisboa
Lisbon
Portugal

lmcarvalho@fc.ul.pt

Jared Caywood
School of Natural Sciences
University of California
Merced, CA
USA

jcaywood4@yahoo.com

Iñaki Comas
Unitat Mixta d'Investigació en Genòmica i Salut
Centre Superior d'Investigació en Salut Pública/UV-Institut Cavanilles de Biodiversitat i Biologia Evolutiva
València
Spain

inaki.comas@uv.es

Vincent Daubin
Laboratoire de Biométrie et Biologie Evolutive
CNRS UMR 5558
Université de Lyon 1
Villeurbanne
France

vincent.daubin@univ-lyon1.fr

Francisco Dionisio
Centro de Biologia Ambiental (CBA) and Departamento de Biologia Vegetal
Faculdade de Ciências da Universidade de Lisboa
Lisbon;
Instituto Gulbenkian de Ciência
Oeiras
Portugal

dionisio@fc.ul.pt

Iolanda Domingues
Centro de Biologia Ambiental (CBA)
Faculdade de Ciências da Universidade de Lisboa
Lisbon
Portugal

iolandalopesdomingues@gmail.com

Joshua Finley
School of Natural Sciences
University of California
Merced, CA
USA

thejoshdude@gmail.com

M. Pilar Francino
Unitat Mixta d'Investigació en Genòmica i Salut
Centre Superior d'Investigació en Salut Pública/UV-Institut Cavanilles de Biodiversitat i Biologia Evolutiva
València
Spain;
School of Natural Sciences
University of California
Merced, CA
USA

francino_pil@gva.es

A. Carolin Frank
School of Natural Sciences
University of California
Merced, CA
USA

cfrank3@ucmerced.edu

Fernando González-Candelas
Unitat Mixta d'Investigació en Genòmica i Salut
Centre Superior d'Investigació en Salut Pública/UV-Institut Cavanilles de Biodiversitat i Biologia Evolutiva
València
Spain

gonzalez@uv.es

Serena Lai
School of Natural Sciences
University of California
Merced, CA
USA

slai5@ucmerced.edu

Florent Lassalle
Laboratoire de Biométrie et Biologie Evolutive
CNRS UMR 5558;
Laboratoire d'Ecologie Microbienne
CNRS UMR5557
Université de Lyon 1
Villeurbanne
France

florent.lassalle@univ-lyon1.fr

Helena Mendes-Soares
Department of Biology
Indiana University
Bloomington, IN
USA

mhmendes@indiana.edu

Sílvia C.M. Mendonça
Centro de Biologia Ambiental (CBA)
Faculdade de Ciências da Universidade de Lisboa
Lisbon;
Instituto Gulbenkian de Ciência
Oeiras
Portugal

silviacm@gmail.com

Bernardino Moreira
Centro de Biologia Ambiental (CBA)
Faculdade de Ciências da Universidade de Lisboa
Lisbon
Portugal

sphynks.bmm@gmail.com

Peter Mullany
Department of Microbial Diseases
UCL Eastman Dental Institute
University College London
London
UK

p.mullany@eastman.ucl.ac.uk

Teresa Nogueira
Centro de Biologia Ambiental (CBA)
Faculdade de Ciências da Universidade de Lisboa
Lisbon;
Centro de Investigação em Saúde e Ambiente
Escola Superior de Tecnologia da Saúde do Porto
Instituto Politécnico do Porto
Vila Nova de Gaia
Portugal

teresainogueira@gmail.com

Ana M. Reis
Departamento de Biologia Vegetal
Center for Biodiversity
Functional & Integrative Genomics (BioFIG)
Faculdade de Ciências da Universidade de Lisboa
Lisbon
Portugal

amreis@fc.ul.pt

Adam P. Roberts
Department of Microbial Diseases
UCL Eastman Dental Institute
University College London
London
UK

adam.roberts@ucl.ac.uk

Chad Swanlund
School of Natural Sciences
University of California
Merced, CA
USA

aquapollo@hotmail.com

This book is dedicated to my parents, Lucien and Marisa

Preface

Horizontal gene transfer (HGT) is probably the most distinctive aspect of microbial evolution. Eukaryotes have evolved genetic systems that are closed to exchanges among organisms from different species. In contrast, the microbial world has produced a vast array of mechanisms that allow the transfer of genes among different organisms and across myriads of mobile genetic elements (MGEs). The result is a world of great diversity and adaptability, in which novelties created in an organism can find their way into a variety of different taxa. In this book, we have aimed at exploring the processes and mechanisms of HGT, their extent and limitations, and the consequences of the openness of microbial genetic systems for different evolutionary processes.

The repercussions of horizontal transfer on gene, genome and organismal evolution are explored across several chapters of the book. In Chapter 1, I address how HGT separates gene evolution from organismal evolution, and the impact of this separation on different classes of genes, whereas, in Chapter 5, Comas and González-Candelas analyse the process of sequence divergence of transferred genes at different evolutionary time-scales. In Chapter 2, Lassalle and Daubin focus instead on evolution at the genome level, setting off from the realization that closely related bacterial genomes can vary enormously in gene repertoires, mainly due to HGT. Chapters 3 and 9 address remarkable examples of how HGT enables organismal adaptation. In Chapter 3, Berglund and Frank describe the role of HGT in the evolution and ecology of microbial host-adaptation, whereas, in Chapter 9, Barlow and her collaborators analyse how the evolution of antibiotic resistance genes is affected by both HGT and intragenic recombination.

The processes and mechanisms of HGT, as well as the conditions that may facilitate or hinder them, are also addressed in detail in several chapters. In Chapter 4, González-Candelas and myself review the barriers that control genetic exchange by different mechanisms, and discuss the multiplicity of processes with the potential to modulate, modify or alleviate such barriers. In Chapter 6, Aminov describes the diversity of MGEs and the frequency of HGT in terrestrial, aquatic and host-associated ecosystems, and discusses the factors that may affect MGE-mediated HGT in these different environments. In addition, Mullany and Roberts discuss in Chapter 8 how the advent of metagenomics is advancing our knowledge of the MGEs that are found in diverse environments, and present the techniques that can be applied to discover new MGEs through this type of studies. In a more theoretical approach, Dionisio and his collaborators investigate in Chapter 7 the forces that ensure plasmid maintenance among bacteria.

Overall, we hope that this book conveys our enthusiasm for HGT research and the need to further investigate the detailed dynamics of HGT processes in nature and their repercussions on the evolution of prokaryotes.

Gene Survival in Emergent Genomes

1

M. Pilar Francino

Abstract

As genome sequences have accumulated for closely related bacteria, our view of bacterial genomes has radically changed. Genome comparisons have demonstrated that little 16S rRNA sequence divergence can be accompanied by large differences in total gene repertoire, and even populations of a single 16S rRNA species are made up of vast numbers of genomic varieties. Much of the observed variation among closely related genomes is attributable to gains and losses of genes that are acquired horizontally, in addition to genomic rearrangements, including gene duplications. The genomic flexibility that results from these mechanisms certainly contributes to the ability of bacteria to survive and adapt in varying environmental challenges. However, the duplicability and transferability of individual genes imply that natural selection should operate, not only at the organismal level, but also at the level of the gene. Therefore, it can be argued that genes are semiautonomous entities capable of responding to natural selection at different levels: selfishly, because they can 'reproduce' by duplication and 'migrate' by Horizontal Gene Transfer (HGT), and cooperatively, because their long-term survival depends on their ability to be replicated as parts of integrated, emerging genomes. The extent of gene autonomy, or the tightness of the association between gene and genome, can greatly vary, mostly depending on the relative strengths of the selective processes acting at each of these two different, but directly related, levels of organization.

Introduction

In the early days of genomics, whole genome sequences were collected for increasing numbers of bacterial species, with little attention paid to the variation existing within species at the genome level. This practice stemmed from diverse circumstances, including the limitations imposed by the cost of genome sequencing, but also the perception of the bacterial genome as a relatively static entity, perhaps akin to the typological species of animals and plants perceived by naturalists before the advent of evolutionary theory. As the cost of large-scale sequencing has decreased, whole genome sequences have begun to accumulate for numerous closely related strains and species of bacteria, radically changing our view of bacterial genomes. Comparisons of sets of closely related bacteria have demonstrated that little 16S rRNA sequence divergence can be accompanied by large differences in total gene repertoire. In some cases, little 16S rRNA divergence can be accompanied as well by considerable changes in the global size, organization and GC content of the genome (Normand *et al.*, 2007); extreme examples are seen in the case of intracellular parasites and symbionts, which suffer rapid genome reduction and degradation (Moran, 1996; Mira *et al.*, 2001). Furthermore, studies of environmental communities have revealed that even populations of a single 16S rRNA species are made up of vast numbers of genomic varieties, differing substantially in genome size, and in the presence, arrangement and copy number of different genes. In fact, the total estimated diversity of genotype structures in single environmental populations may largely exceed allelic diversity at individual gene loci (Thompson *et al.*, 2005).

Much of the observed variation among closely related genomes is attributable to gains and losses of genes that are acquired horizontally (Daubin and Ochman, 2004; Coleman *et al.*, 2006). In particular, mobile genetic elements (MGEs), which include transposons, phages and plasmids, can contribute the majority of the gene content differences among closely related strains (Hayashi *et al.*, 2001; Perna *et al.*, 2001). In addition, genomic rearrangements, including large duplications, deletions and inversions, are detected at high frequencies during growth of clonal populations in the laboratory, and are also likely to contribute widely to natural intra- and interspecific variation (Reams *et al.*, 2010). The genomic flexibility that results from the variety of mechanisms just mentioned must certainly contribute to the ability of bacteria to survive and adapt in varying environmental challenges. However, limiting ourselves to an organism-centred view misses the fact that, if genes have the capacity to proliferate within genomes by duplication and to spread across them by HGT, natural selection will also operate at the gene level, on one hand, and at the level of the communities of organisms that are exchanging genes, on the other. This should force us to broaden our conceptions of what the units of evolution are and how they behave in the prokaryotic world. It has long been recognized that MGEs can evolve selfishly. However, because of HGT, the distinction between selfish, mobile DNA and the sedentary host genome is blurry. It can be argued that all genes are entities capable of responding to natural selection at different levels: selfishly, because they can 'reproduce' by duplication and 'migrate' by intragenomic translocation or horizontal transfer, and cooperatively, because their long-term survival depends on their ability to be replicated as parts of larger integrated genetic units. Genes can thus be compared to organismal species within ecological communities: semiautonomous entities that can only function within the larger context of a higher level of organization, and whose dynamics will be regulated by interactions with the other community members and their common environment. Natural selection at the higher level of organization, the organismal level, will dictate the combinations of genes that are fit and will regulate their functional interactions.

Genes as semiautonomous entities

Given the evolutionary constraints imposed at the organismal level and the lack of the specialized mechanisms of copy number increase and transfer that are available to MGEs, most genes will need to rely on their fitness benefits to the organism in order to persist. Nevertheless, most genes appear to display certain autonomy in their evolutionary paths, although the extent of this autonomy can greatly vary. What processes account for such variability? Although molecular mechanisms will certainly be involved in influencing the likelihood of a gene to duplicate or to transfer between genomes, it is likely that the main causes lye in the relative strengths of selective processes acting at different levels. In the following sections, I will explore the factors that may affect the differential propensity of genes to effectively proliferate within and among genomes.

Gene copy number increase

Both genome sequence comparisons and laboratory experiments indicate that genetic duplications and larger gene amplifications are common events. In fact, tandem duplications are probably the most common mutation type in bacterial cultures, in terms of their rate of formation and their frequency (Reams *et al.*, 2010). Recent experimental work has estimated that duplications of a given locus occur at rates of 10^{-3}–10^{-6} per cell division, i.e. 10^{3}–10^{6} times more often than point mutations. Moreover, after gene duplication, further copy number increases can occur at a high rate (in the order of 10^{-2} per cell division). As a result, and in spite of fitness costs and loss of duplicates through homologous recombination (HR), the steady-state frequency of duplications is such that long-term bacterial populations can be considered stably polymorphic for duplications of every region of their genome. For instance, roughly 10% of cells in an unselected overnight culture of *Salmonella enterica* carry a duplication of some unspecified chromosomal region, including a small proportion of large gene amplifications (Anderson and Roth, 1977; Reams *et al.*, 2010). Genome sequence analyses corroborate that genomes in all domains of life are often full of gene duplicates. Many of the available complete genome sequences of bacteria, archaea and eukaryotes have revealed large numbers of gene duplications and lineage-specific gene expansions (Jordan *et al.*, 2001; Conant and Wagner, 2002; Lespinet *et al.*, 2002; Francino, 2005).

Duplicability: selective constraints

Genome sequence and experimental studies have also shown that the observed frequency of duplications varies greatly among genes. The percentage of cells in a *S. enterica* culture carrying a duplication of a specified gene may range widely between 0.005% and 3% (Anderson and Roth, 1977). Similarly, computational analyses across numerous organisms have shown that the duplicability, or tendency of a gene family to proliferate within a genome, is highly variable (Conant and Wagner, 2002; Hooper and Berg, 2003a,b; Papp *et al.*, 2003; Yang *et al.*, 2003; Davis and Petrov, 2004; Marland *et al.*, 2004; Prachumwat and Li, 2006). Some genes are nearly universally found in single copy across all genomes (Conant and Wagner, 2002; Ciccarelli *et al.*, 2006; Prachumwat and Li, 2006; D'Antonio and Ciccarelli, 2011), others tend to form small gene families, and some appear greatly amplified in diverse organisms (Kondrashov *et al.*, 2002; Lespinet *et al.*, 2002; Gevers *et al.*, 2004; Francino, 2005).

Although location in the bacterial genome can affect the rate of duplication of DNA sequences (Reams *et al.*, 2010), the observed variation in duplicability for different loci is likely to be most influenced by the fitness effects of the corresponding duplications. In accordance, a relationship between gene function and duplicability has been noted for numerous organisms, with ribosomal genes and transcription factors often standing out as the most refractory to gene duplication (Jordan *et al.*, 2001; Conant and Wagner, 2002; Kondrashov *et al.*, 2002; Papp *et al.*, 2003; Marland *et al.*, 2004; Francino, 2005; Prachumwat and Li, 2006; Sorek *et al.*, 2007; D'Antonio and Ciccarelli, 2011). Aiming at rationalizing these duplicability differences, Papp *et al.* (2003) proposed the dosage balance hypothesis, which postulates that genes coding for subunits of protein complexes tend to have a lower duplicability, because of their high potential to cause dosage imbalances. Their proposition was based on analyses of the yeast *Saccharomyces cerevisiae* genome and, since then, other computational analyses have corroborated that gene duplicability in yeast indeed decreases with the number of subunits in a protein, or protein complexity (Yang *et al.*, 2003), and have shown that duplicability is most constrained for proteins that form hetero-complexes (complexes made up of different subunits; Lin *et al.*, 2007). In *Escherichia coli*, however, computational analyses have failed to show a general correlation between protein complexity and duplicability (Wellner *et al.*, 2007).

Other factors have been associated with gene duplicability. Protein connectivity, a property conceptually related to protein complexity and defined as the number of interactions in protein–protein networks, has been shown to restrict duplicability in yeast and in *E. coli*, suggesting that dosage imbalance can perturb not only the operation of molecular complexes but also network equilibria (Li *et al.*, 2006; Prachumwat and Li, 2006; Wellner *et al.*, 2007; D'Antonio and Ciccarelli, 2011). The position of a protein in interaction networks can also affect duplicability, independently of the actual number of interactions. For instance, network centrality, measured as the number of shortest paths that pass through a given protein, characterizes how essential a protein is in maintaining communication between the network nodes, and has been found to restrict duplicability (D'Antonio and Ciccarelli, 2011). Analyses involving several yeast species have revealed that genes that impose a high fitness cost when deleted are also rarely duplicated, indicating a selective pressure for these genes to be retained but maintained in a single copy (He and Zhang, 2006). Similarly, the majority of genes that are present across all fungal genomes or that have been experimentally shown to be essential in yeast are nearly always found in single copy (Waterhouse *et al.*, 2011). In addition, these universal single-copy fungal genes have slower sequence evolution rates than the universal fungal genes that are found in multiple copies (Waterhouse *et al.*, 2011). In contrast, analyses not restricted to universal genes suggest that genes with slower rates of sequence evolution more often generate duplicates in yeast and in bacteria (Davis and Petrov, 2004; Jordan *et al.*, 2004), indicating that the relative strength of duplicability and sequence change constraints may vary among genes. Finally, and most significantly, genes that originated during early periods in the evolution of life, such as those present in LUCA, the last universal common ancestor of bacteria, archaea and eukaryotes, are most often found in single copy (D'Antonio and Ciccarelli, 2011).

Duplicability: positive selection and diversification

Although the fitness effects of most duplications, as those of other mutations, will likely be deleterious or neutral, there is ample evidence that gene amplification and increased production of the encoded protein can occasionally provide specific selective advantages. Adaptation by means of gene duplications and larger amplifications has been repeatedly documented in bacteria and in yeast. Gene amplifications have been implicated in enhanced virulence for some bacterial pathogens, as well as in increased production or fixation of host-required nutrients in symbionts. Gene amplification has also been demonstrated to underlie instances of bacterial resistance to antibiotics and heavy metals, as well as experimental adaptation to growth at high temperature and on limiting or unusual carbon sources (Romero and Palacios, 1997; Riehle *et al.*, 2001; Dunham *et al.*, 2002; Reams and Neidle, 2003; Zhong *et al.*, 2004).

However, it seems unlikely that a series of identical genes would be maintained in long-term evolution for the purpose of increasing the amount of gene product. To that effect, alternative, and possibly more efficient, strategies would be available to an organism, such as increases in the expression of one or a few gene copies through regulatory changes. Large, 'clonal' gene amplifications are therefore likely to be transient in nature. Accordingly, analyses of the distribution of gene family sizes across different genomes have detected an excess of very large gene families that are often present in shallow lineages, indicating that they have appeared only recently (Gerstein, 1997; Huynen and van Nimwegen, 1998; Qian *et al.*, 2001; Harrison and Gerstein, 2002; Lespinet *et al.*, 2002; Karev *et al.*, 2003; Francino, 2005). Given that deletion rates are high in most bacteria (Mira *et al.*, 2001; Gomez-Valero *et al.*, 2007), genetic drift should favour the rapid elimination of gene copies if they become superfluous owing to the appearance of better adapted coding sequence or regulatory variants. Most of the gene clones can therefore be expected to eventually disappear, unless they're rescued by adaptive mutations that allow them to occupy a novel functional niche. If there is formation of new, separate niches, genic lineage diversification will occur. Eventually, neutral and adaptive sequence divergence, possibly reinforced by selection against recombinant sequences, will generate paralogous genes adapted to specific functions and evolving in different paths. In the long term, families and superfamilies of paralogous genes can be generated by this diversification process (Francino, 2005).

Gene mobility and transfer

Computational and phylogenetic analyses of bacterial genome sequences have documented instances of horizontal transfer of all types of genes and at all phylogenetic distances (Gogarten *et al.*, 2002). However, as for duplicability, the capacity of a gene to transfer horizontally will likely depend on numerous factors. These should include the possibility of rising, at least occasionally, to high copy numbers. If a gene is present in numerous copies in a large number of genomes, it will be more likely to randomly associate with a MGE that might transfer it to another organism. Also, high copy number will increase the likelihood that one copy might withstand the vagaries of surviving as free DNA in the environment after cell death, or of undergoing a successful recombination event with a host chromosome after transformation.

Other factors that increase the likelihood of association with MGEs will also necessarily favour the horizontal spread of otherwise sedentary genes. MGEs can constitute a significant portion of many bacterial genomes, representing, for instance, 1% of the genome in *E. coli* and nearly one-quarter of the genome of *Enterococcus faecalis*. Importantly, the distribution of MGEs across the genome is not homogeneous. In several genomes, including *E. coli* and *Bacillus subtilis*, the replication terminus region is enriched in prophage insertions, suggesting that genes located in this region of the chromosome could be transferred more often by transduction upon prophage excision. On the other hand, a comprehensive analysis of *E. coli* chromosomal sequences associated with lytic cycle phage Mu virions indicated that it was rather those genes encoded near the origin that were overrepresented (Manna *et al.*, 2004). This likely reflects the fact that, in rapidly growing bacteria, replication reinitiates before the previous round has been completed and therefore sequences near the origin are found in a higher copy number per cell (Li *et al.*, 2003). An even stronger bias was detected in relation to transcription, which was identified as a powerful inhibitor of phage insertion, presumably owing to interference between RNA polymerase and phage integration enzymes (Manna *et al.*, 2004).

Transferability: selective constraints

Although copy number, genome location and other factors may influence the likelihood that a gene will find itself in a MGE or will otherwise find its way into another organism, the fitness effects of the gene on the receiving host are likely to be the main determinants of its transferability. A relationship between protein function and HGT has been revealed by numerous computational analyses of bacterial genomes, which have clearly established a low level of HGT for informational molecules, such as those related to transcription and translation (Jain *et al.*, 1999; Brown *et al.*, 2001; Brochier *et al.*, 2002; Nakamura *et al.*, 2004; Ciccarelli *et al.*, 2006). This relationship prompted the proposal of the 'complexity hypothesis', which states that proteins involved in many interactions are less likely to function properly in a new host after transfer (Jain *et al.*, 1999). Explicit computational tests of this hypothesis have demonstrated that genes encoding proteins with multiple interactions are indeed less likely to have been involved in horizontal transfers (Wellner *et al.*, 2007; Davids and Zhang, 2008; Lercher and Pal, 2008). The reasons behind this observation could be several. Dosage imbalance following transfer, as proposed for duplicability, could interfere with the operation of protein complexes and protein interaction networks; and/or proteins necessitating multiple interactions to function could fail to establish them in a different host and therefore provide no selectable functionality, or have detrimental effects by competing with the native homologues and resulting in dysfunctional interactions.

Sorek *et al.* (2007) addressed transferability differences across genes in an experimental context, by taking advantage of clone libraries prepared for shotgun genome sequencing projects. Using this resource, they analysed the ability of the *E.coli* DH10B cloning host to survive in the presence of plasmids containing different genomic regions representing the entire genomes of 79 bacterial species. This procedure identified genes that consistently failed to transfer from numerous donors. Moreover, such genes were also estimated to undergo little HGT in a phylogeny-based computational analysis of 191 sequenced genomes. Functionally, the largest fraction of non-transferable genes belonged to COG (Clusters of Orthologous Genes) classes involved in translation, followed by those involved in transcription and the biogenesis of cell walls and membranes. Clearly, these gene classes encode proteins that participate in complex, central processes of the cell, requiring many molecular interactions. These results therefore represent an experimental validation of the complexity hypothesis for HGT; moreover, many of the non-transferable genes failed to transfer even when the donor was a closely related strain of *E. coli*, and were also found to be universal single-copy genes (SiCo), never duplicated across bacterial genomes, indicating that dosage imbalance was likely to be one of the main reasons behind their lack of transfer. In accordance with this notion, further experiments involving cloning of non-transferable ribosomal protein genes under inducible promoters demonstrated that the deleterious effects of increased gene expression were a predominant cause of transfer failure. In contrast, other experiments that have addressed the effects of cloning foreign genes encoding subunits of large protein complexes have not detected detrimental fitness effects (Wellner and Gophna, 2008; Omer *et al.*, 2010). Interestingly, the genes assayed in these experiments (*accA*, which encodes the alpha subunit of carboxyl-transferase within the acetyl-CoA carboxylase of fatty acid synthesis and *rpoB*, which encodes the beta subunit of RNA polymerase) were not detected to be non-transferable in the analyses of Sorek *et al.* (2007), suggesting that the effects of expressing a foreign gene may differ substantially even when only genes encoding for subunits of large protein complexes are considered.

Transferability: advantageous effects of transferred functions

Clearly, the permanence of a gene in a new genome after transfer will be facilitated if the gene confers a valuable adaptation to the organism. Given stable conditions, bacterial populations are likely to be well adapted to their environment, thanks to their large sizes, short generation times and elevated mutation rates that allow them to explore the fitness landscape in relatively short amounts of time. However, conditions in nature are seldom temporarily stable or spatially uniform, allowing for frequent opportunities to improve the match between genome and environment. Such challenges are often met by HGT. For example, environmental distributions of toxic compounds, such as antibiotics or heavy metals, tend to be spatially and temporarily variable (Eberhard, 1990) and the genes that confer resistance to them are often encoded in plasmids, genomic islands and other MGEs undergoing frequent transfer (Levin and Cornejo, 2009; Sobecky and Coombs, 2009; Rankin *et al.*, 2011). Further, antagonistic biotic interactions, such as those between pathogenic bacteria and their animal or plant hosts or those between bacteria and their phages, can lead to arms races that permanently favour diversity (Van Valen, 1974). Numerous studies have documented the involvement of HGT in generating diversity at bacterial loci involved in host infectivity, evasion of host defence systems or phage resistance (Smith and Selander, 1991; Feavers *et al.*, 1992; Hall and Sharp, 1992; Marklund *et al.*, 1992; Spratt *et al.*, 1992; Harbaugh *et al.*, 1993; Murray *et al.*, 1993; Li *et al.*, 1994; Wang *et al.*, 2003; Ma *et al.*, 2006; Patil *et al.*, 2007; Avrani *et al.*, 2011).

Genes involved in interactions among bacteria, both cooperative and antagonistic, are also often encoded in MGEs or show other hallmarks of horizontal transfer (Livermore, 1995; Nogueira *et al.*, 2009; Riley and Wertz, 2002; Rankin *et al.*, 2011). It is interesting to note that, when a gene or MGE transfers to a different individual, not only the recipient gene community is altered, but also the relatedness between the two individuals that have exchanged genetic material. By increasing relatedness, HGT could favour the evolution of cooperation among gene-exchanging individuals, and it has recently been proposed that MGEs promote the emergence of cooperative social traits (Nogueira *et al.*, 2009; Rankin *et al.*, 2011). Importantly, in prokaryotes, relatedness between individuals will vary substantially across loci, particularly between MGEs and the rest of the genome, raising the possibility of conflict regarding what type of interaction with a given neighbour might be most advantageous.

Adaptation to novel environments or changing environmental conditions has also been repeatedly documented to occur through acquisition of horizontally transferred genes (Lawrence, 1997, 1999; Gogarten *et al.*, 2002; Coleman *et al.*, 2006; Alcaraz *et al.*, 2008; Penn *et al.*, 2009; Wiedenbeck and Cohan, 2011; Wu *et al.*, 2011). Similarly, the adoption of novel lifestyles, such as symbiosis or pathogenesis and their diversification in terms of host range and tropism, is most often enabled by the acquisition of MGEs (Ochman and Groisman, 1996; Boyd and Hartl, 1998; Hacker and Kaper, 2000; Ochman *et al.*, 2000; Gogarten *et al.*, 2002; Toth *et al.*, 2006; Stinear *et al.*, 2007; Stinear *et al.*, 2008; Berglund *et al.*, 2009; Marchetti *et al.*, 2010; Schmidt, 2010; Silby *et al.*, 2011). If new niches persist, bacterial populations that have gained access to them by HGT may in time become independent lineages with their own evolutionary paths, and thus HGT can be considered a motor of prokaryotic speciation and long-term diversification (Lawrence, 1997,1999, 2002; de la Cruz and Davies, 2000; Cohan, 2001; Gogarten *et al.*, 2002). Amazing examples of appearance of major lineages with specific biologies introduced by HGT include the emergence of bacterial methanotrophs via acquisition of archaeal genes (Chistoserdova *et al.*, 1998) and that of cyanobacteria via the gain of a second photosystem allowing for oxygenic photosynthesis, possibly transferred from Firmicutes (Xiong *et al.*, 1998).

Thus, given that HGT seems most valuable for recipient cells when it serves to generate adaptive novelty and open new niches, it is clear that certain functions will have specific properties that will make them most likely to be advantageous after transfer. Functions that are basically conserved across organisms, such as those encoded by housekeeping genes, are unlikely to result in a selective advantage to the recipient cell. Their probable fate after transfer should be loss from the population through genetic drift (Boyd *et al.*, 1994). On the other end of the spectrum, functions that are as niche-specific as to be of no use in any other environment will also be extremely limited in their transferability. Intuitively, functions which are not universally required but which may be useful in a variety of niches, even if spatially or temporally restricted, should be most prone to spread by HGT.

Transferability: factors that facilitate functionality in diverse backgrounds

In addition to the specific function conferred by a gene, other operational factors may affect the likelihood that it will successfully perform in a new genomic, cellular or environmental background. The potential barriers that may impede the integration and expression of a horizontally transferred gene in a new genome are many, and they will be discussed in Chapter 4. Here we will focus on exploring gene and protein properties that may facilitate their operability across a variety of organisms.

Numerous aspects of gene expression are highly tuned to the particular sequence characteristics of each genome. Genes of exogenous origin are likely to have different GC content and codon usage than ancestral genes, which may compromise their expression in several ways. Many bacteria possess abundant nucleoid-associated proteins that modulate DNA conformation and contribute to the regulation of DNA metabolic processes such as replication, recombination, repair and transcription. The histone-like nucleoid-structuring protein (H-NS) is one of the major protein components of the nucleoid structure and binds large regions of DNA, down-regulating gene expression, with a marked preference for AT-rich sequences. In general, horizontally transferred sequences tend to be AT-rich. Consequently, many horizontally transferred sequences are recognized by H-NS and transcriptionally repressed, in a process termed xenogeneic silencing (Lucchini *et al.*, 2006; Navarre *et al.*, 2006, 2007). This repression may play a crucial role in protecting the cell from potentially detrimental effects of foreign sequences, particularly phages, which are significantly richer in AT than their host (Rocha and Danchin, 2002). However, in the long term, xenogeneic silencing may facilitate the acquisition of AT-rich genes, by initially masking their effects while allowing their retention, thus enabling the cell to build up cryptic capabilities that may become advantageous if expressed in a regulated manner. Over time, regulated expression of foreign genes may come about by means of the evolution of specific DNA–protein interactions that may outcompete H-NS and render the DNA accessible to the transcriptional apparatus of the cell (Navarre *et al.*, 2007). In accordance with this scenario, several *Salmonella* pathogenicity islands become overexpressed upon removal of H-NS from the cell, with deleterious consequences for bacterial fitness (Lucchini *et al.*, 2006). The likelihood of eventual expression of foreign genes may also be influenced by their GC content because RNA-polymerase and transcription factors are thought to preferentially bind AT-rich motifs across bacterial taxa (Huerta *et al.*, 2006; Froula and Francino, 2007; Cordero and Hogeweg, 2009). Therefore AT-rich sequences may be more likely to carry or evolve motifs that can promote gene expression in different genomic backgrounds.

Given the series of constraints that may dampen the expression of exogenous sequences, coupled to the fact that increased amounts of gene product may be needed in heterologous contexts to compensate for suboptimal function, a reduced expression level is likely to be an important barrier to gene exchange. In accordance, replacements of ribosomal genes, the products of which are required in high amounts, with orthologues from different species result in a significant decrease in fitness that can be partially restored if the transferred genes are duplicated (Lind *et al.*, 2010). Computational analyses have also shown that the occurrence of gene duplications is increased among horizontally transferred genes (Hooper and Berg, 2003a). The potential repercussions of low expression levels should be less dramatic for genes whose products are not required in high amounts in order to confer a selective advantage to the recipient cell, so that these genes should transfer more easily across different genomes. Moreover, the most highly expressed genes in a genome are the ones that use the most restricted sets of codons, because selection on translational efficiency is strongest on these genes and favours the usage of codons that correspond to the most abundant tRNAs in the cell, which vary across species (Ikemura, 1985; Sharp and Li, 1986; Kanaya *et al.*, 1999; Rocha, 2004; Tuller, 2011). In contrast, genes that are not highly expressed have a less biased codon usage and their expression may be less affected by differences in tRNA pools among organisms. Accordingly, in *E. coli*, horizontally transferred genes have lower levels of expression (Davids and Zhang, 2008), and analyses of HGT events among 190 prokaryotes show that genes whose translation efficiency is less sensitive to changes in the tRNA pool have been transferred more often (Tuller, 2011).

At the protein level, it is logical to suppose that proteins that have a more stable structure will be more tolerant to the different intracellular conditions and mutational backgrounds that may be encountered in a new organism, and/or will be able to accommodate the changes required to withstand these new circumstances. It is known from experimental evolution studies that proteins that are thermostable can withstand the introduction of large numbers of mutations and still maintain their function (Besenmatter *et al.*, 2007). Further, the mutational robustness conferred by stability can facilitate evolvability, by enabling the protein to accept a wider range of beneficial mutations while still folding to its native structure and/or retaining catalytic function (Wagner, 2005; Bloom *et al.*, 2006). Other proteins may be evolvable by virtue of other structural properties, such as conformational plasticity or modularity (Tokuriki and Tawfik, 2009; Yadid *et al.*, 2010). Evolvability and functional versatility should broaden the range of organisms into which a gene may transfer, by enabling the gene to confer related but distinct selectable phenotypes in organisms that are distantly related or have different environments or lifestyles. Horizontally transferred proteins are known to undergo shifts in function to better serve the specific needs of their new hosts. As an example, omptins are a family of outer membrane proteases that have spread horizontally through a large variety of Gram-negative bacteria that infect vertebrates and plants, and their functions have adapted to the particular lifestyles of their different hosts. Moreover, site-specific mutagenesis experiments have succeeded in changing the function of a plant pathogen omptin to resemble that of its close homologue, the plasminogen activator of *Yersinia pestis,* so that the modified protein has an altered protease activity and confers to *E. coli* the capacity to invade human cells. In this case, the mutational robustness and evolvability of omptins may ensue from their sturdy membrane-embedded β-barrel structure combined with flexible surface loops (Haiko *et al.*, 2011). A similarly amazing example of functional versatility and evolvability is provided by another type of bacterial membrane structures that have undergone extensive HGT, the type IV pili. Type IV pili are long, flexible filaments, formed by polymerization of pilin subunits, that extend from the bacterial surface and can be found in Gram-negative and Gram-positive bacteria. These versatile fibres have a unique ability to promote multiple and strikingly different phenotypes that contribute to cell survival and pathogenesis, such as attachment to surfaces, aggregation, uptake of DNA during transformation, conjugation, protein secretion, motility, host colonization and biofilm development. No other pili are found in as many or as diverse bacteria as type IV pili, probably by virtue of their functional versatility (Wall and Kaiser, 1999; Hansen and Forest, 2006; Pelicic, 2008; Filloux, 2010).

Another operational advantage when it comes to HGT may be the capacity to function as a single protein, so that dependency on the recipient organism is reduced. As already discussed above, proteins that have fewer interactions are indeed more often involved in HGT (Jain *et al.*, 1999; Wellner *et al.*, 2007; Davids and Zhang, 2008; Lercher and Pal, 2008). However, it is not clear whether this pattern is mainly caused by a lack of positive selection for proteins that cannot function without their coevolved partners, or, rather, by negative selection due to dosage imbalances and dysfunctional interactions among exogenous and native proteins. Successful HGT should also be facilitated for groups of functionally related proteins that are encoded in a compact genetic module that can spread as a single unit. The idea that physical proximity of all the genes required for a selectable phenotype enhances the probability of horizontal co-transfer is the basis of the selfish operon model (Lawrence and Roth, 1996; Lawrence, 1999). This model postulates that the organization of bacterial genes into operons is mostly beneficial to the constituent genes, rather than to the cell that carries them, as it facilitates their successful spread across organisms by horizontal transfer, and that it is this selfish selective advantage that is the main driver of operon evolution. However, operon structures are abundant both for genes that are thought to have undergone HGT and for essential, ancestral SiCo genes (Pal and Hurst, 2004; Price *et al.*, 2005; Fang *et al.*, 2008). Therefore, it is likely that HGT is not the only force behind operon formation and that processes that affect organismal fitness, such as selection for reducing the amount of regulatory information required to optimize co-expression and/or selection for minimizing the proportion of deletions that may hit essential genes, also contribute to the evolution of this type of gene organization (Omelchenko *et al.*, 2003; Price *et al.*, 2005; Fang *et al.*, 2008). Nevertheless, once formed, operons will facilitate the transfer of complex functions that require several proteins, and cases of horizontal transfer of entire operons are clearly abundant.

Clustering of functionally related genes such that complex functions can spread across organisms is also commonly found in MGEs. Transposons and plasmids can carry complex gene clusters, for instance, those required for the utilization of novel carbon sources or for the degradation of xenobiotic compounds. Genomic islands (GIs), prophages and other large genomic regions probably introduced by HGT often consist of different blocks of genes representing functional units involved in host–environment interactions, metabolism, intracellular or intercellular DNA mobility, drug resistance or antibiotic synthesis, and that are present or absent as entire modules from closely related genomes (Ogier *et al.*, 2010).

Transferability: selfish measures to ensure permanence

The capacity of genes to behave selfishly is probably best exemplified by systems that act as addiction modules. Numerous examples of such systems have been noted on bacterial chromosomes or extrachromosomal replicons, where they can remain for long periods of time (Lawrence, 1999). They consist of an operon containing two genes, which act as a toxin and an antitoxin for the cell that carries them. The toxin is a protein that kills cells if expressed above a certain level, and the antitoxin, which can be a different protein or an antisense RNA, inactivates the toxin and/or regulates its expression, thereby preventing cell killing. The toxin is more stable and long-lasting than the antitoxin, so that constant (over) production of antitoxin is required for cell survival. Comparative genome analyses show that toxin–antitoxin (T–A) systems have extensive horizontal mobility and are commonly spread via plasmid-mediated HGT (Makarova *et al.*, 2009). In addition, toxins and antitoxins exhibit a certain degree of autonomous evolution, as they can reassort into different T–A combinations, or engage in regulatory interactions *in trans*. For instance, it has been recently demonstrated that a chromosomally encoded antitoxin is capable of inhibiting a toxin encoded in a plasmid (Saavedra De Bast *et al.*, 2008). Furthermore, it has been proposed that the proteins involved in T–A systems may also have alternative cellular functions, and, in particular, that antitoxin proteins may often participate as transcriptional regulators in various forms of stress responses (Makarova *et al.*, 2009).

Other systems share with the T–A addiction modules the poison–antidote principle, such as the DNA restriction–modification (R–M) systems that are ubiquitous in eubacteria and archaea. Most of these systems comprise a DNA methyltransferase that enables recognition of 'self' DNA by methylation of specific nucleotides within particular DNA sequences and a restriction endonuclease that will cleave foreign unmodified DNA. In this case, loss of the modification genes could result in degradation of the bacterial chromosome by the remaining restriction endonuclease. Although R–M systems undoubtedly offer bacteria an important means of protection against invasion by infectious agents, it is clear that these systems also have the potential for selfish behaviour (Naito *et al.*, 1995; Kobayashi, 2001). Like T–A systems, they have undergone extensive HGT and they are commonly linked with plasmids, phages, transposons or integrons.

It is likely that addiction modules, including T–A and R–M systems, depend on both their selfish properties and their benefits to organismal fitness in order to spread and be maintained in bacterial populations (Mochizuki *et al.*, 2006), and therefore they represent a complex example of the capacity of genes to engage in semiautonomous evolutionary paths.

A gamut of strategies for gene survival

We have just seen that large differences in duplicability and transferability exist across genes. At the organismal level, theoretical and experimental evidence has shown that the spread of a symbiont depends on the interplay between its transmissibility and its effect on the fitness of the host, with selection favouring the combination of these factors that maximizes symbiont increase under specific circumstances (Lenski and May, 1994). This combination will determine the type of relationship that develops between a symbiont and its host. The two ends of the spectrum are occupied by highly virulent and highly transmissible parasites, and by host-restricted beneficial symbionts. Conditions that favour transmissibility, such as high host density, select for symbionts with high growth rates that can compensate their burden on individual host fitness by effective transmission to neighbouring hosts. If transmissibility is high enough, highly virulent parasites may survive during long periods of time. Alternatively, if the transmissibility of the symbiont is low, its survival will be linked to that of the individual host that carries it, and selection will favour parasites with decreased virulence and mutualists that confer larger benefits to their hosts. These principles have also been shown to apply to the evolution of MGEs such as plasmids (Stewart and Levin, 1977). At the genic level, there may be a similar gamut of long-term strategies for gene survival, representing different combinations of duplicability and transferability, on one hand, and of reliance on vertical transmission and organismal fitness, on the other.

Genes of the essential core

At one end of the spectrum, genes that are essential for life may entirely rely on their permanence through generations via vertical inheritance. Further, their chances of survival should not increase substantially with HGT, because they are already represented in every living cell and no organism represents an empty niche into which to expand. As mentioned above, their transfer is likely to result in no selective advantage to the recipient cell and to be followed by loss from the population through genetic drift or purifying selection. Therefore, universally essential genes should only respond to selective pressures to increase the fitness of the specific organisms in which they reside, allowing for extensive host specialization in spite of their ubiquity. Their long-term permanence within specific genomes should also enable a high degree of co-adaptation with other stable members of their genomic community, favouring the evolution of a well-integrated core of essential genes. Overall, this universally essential gene core would likely evolve under strong purifying selection for the maintenance of stable, host-specialized and co-adapted functionalities. This mode of evolution should demand evolutionary conservation at numerous different levels, by simultaneously penalizing gene loss, accelerated sequence evolution rates, horizontal transfer and, for highly interactive proteins, gene duplication. From the point of view of phylogenetic enquiry, such constraints signify that genes of the universally essential core represent ideal markers, because their evolutionary dynamics are free from the different processes that commonly shift gene genealogies away from organismal phylogeny. Representative genes of this evolutionary mode are those involved in transcription and translation, which are universally essential, rarely transferred, conservative in sequence and function, generally maintained in single copy and encode highly interactive co-adapted protein complexes (Jain *et al.*, 1999; Brown *et al.*, 2001; Brochier *et al.*, 2002; Nakamura *et al.*, 2004; Ciccarelli *et al.*, 2006; Sorek *et al.*, 2007).

Genes of phylogenetically or ecologically restricted cores

Beyond the core of genes essential for life, many genes may be shared within specific phylogenetic clades or within groups of organisms with similar ecological needs. Indeed, the numbers and types of genes that are conserved increase significantly within progressively shallower phylogenetic lineages (Charlebois and Doolittle, 2004). For instance, whereas all prokaryotes almost certainly share less than 50 genes, over 100 genes are common to all Bacteria (Santos and Ochman, 2004) and Lerat *et al.* (2003) identified 205 single copy genes present across all the gamma-proteobacteria, a large and ancient group that originated over 500 MYA. The gamma-proteobacterial gene core was not restricted to the informational category, as it also included many operational genes. Remarkably, nearly all (203) of these genes generated phylogenies that were concordant, and their concatenation produced a fully resolved and well-supported tree for this proteobacterial class. At shallower phylogenetic levels, the number of shared genes increases and often reaches 1000 within bacterial species. Within the enteric family of the gamma-proteobacteria, the core genomes of the large and well-characterized species *E. coli* and *S. enterica* have been estimated at around 1000 and 2800 genes, respectively (Lukjancenko *et al.*, 2010; Jacobsen *et al.*, 2011). Besides informational genes, these species-level cores mainly include genes involved in the biosynthesis of amino acids, nucleotides, co-factors and proteins as well as in the metabolism of DNA, fatty acids, and phospholipids (Touchon *et al.*, 2009).

Genes that are ubiquitous within restricted phylogenetic clades, such as species or genera, may undergo frequent horizontal shuffling within these groups, depending on the population structure and level of ecological divergence of the different lineages, the involvement of a particular core gene in niche specialization, and the capacity of the organisms to incorporate genes by HR. For instance, phylogenetic analysis of the core genome of the genus *Listeria* revealed that it contained both genes evolving within specific lineages and under niche-specific selection pressures and genes that appeared to have been recombining freely across ecologically divergent lineages and that produced incongruent genus-wide phylogenies (Dunn *et al.*, 2009). In contrast, the core genomes of other genera have been shown to be more recalcitrant to interspecific exchange (Vinuesa *et al.*, 2005). Interestingly, essential and non-essential genes within the *Listeria* core had similar evolutionary dynamics, with no differences in their likelihoods of being exchanged or of undergoing niche-specific selection. Rather, non-random associations were found between niche-specific selection pressure and other high level characteristics, such as metabolic modularity or variability in gene expression.

Non-core accessory genes

A large fraction of the genes unveiled by genomic sequencing projects has shown to be highly variable in terms of their presence and copy number across different genomes, and to be often transferred among organisms with varying degrees of relatedness. Even within a single bacterial species, those genes that are not present in the genome of every intraspecific lineage or strain can amount to a large proportion of the total gene repertoire detected across the species, or pangenome (Tettelin *et al.*, 2005; Mira *et al.*, 2010). The fraction of the pangenome that is not common to every member of the species has been termed the dispensable or accessory genome (Tettelin *et al.*, 2005). Clearly, such dispensability implies that not all members of a species require all of the genes available to it as a whole, or at least not all of the time. Such scenario can be interpreted through two main models: (1) each genomic variant within a species represents a stable evolutionary unit that has adapted to a differentiated, long-lived ecological niche (an ecotype; Cohan, 2006), or (2) genomic variants are labile states that appear by genetic drift or that are selected by transient conditions in a variable environment. In (1), accessory genes will survive in the same genome for as long as the specific conditions that define the ecotype's niche are maintained. In contrast, in (2), accessory genes may survive in the long term via transient associations with different genomes (Fig. 1.1).

Accessory genes, and particularly GIs and other MGEs, that are clearly associated with well-defined niches or lifestyles can certainly be singled out, most notably in the case of pathogens and symbionts. Among free-living bacteria, stable, phenotypically distinct

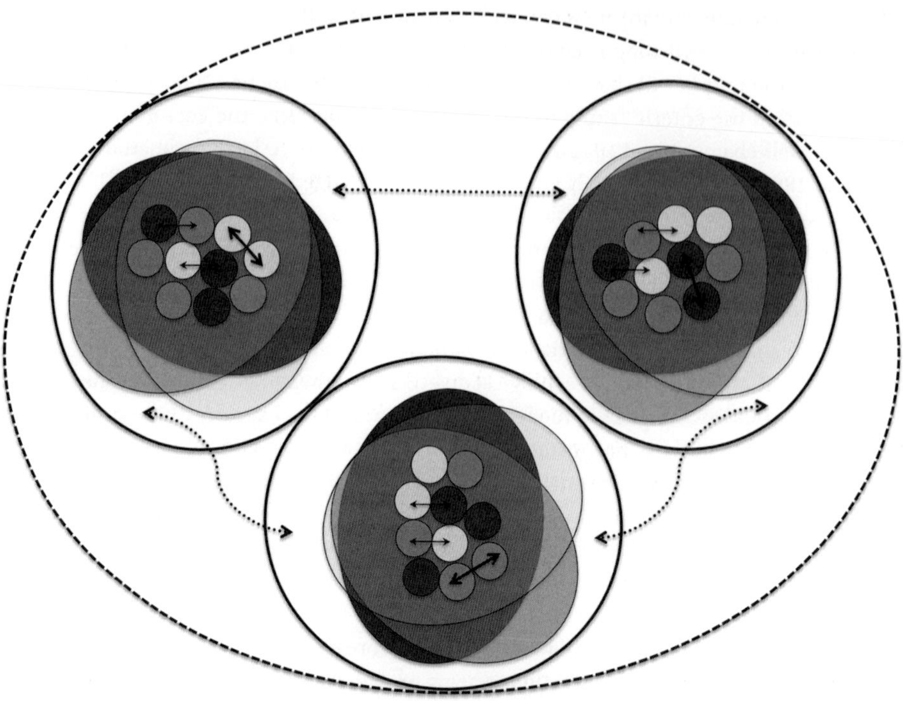

Figure 1.1 Movement of accessory genes within a metacommunity. This hypothetical metacommunity is composed of three local communities connected by organismal dispersal (dotted arrows). Within local communities, genes can be exchanged between organisms (small circles) of the same population (light, medium or dark grey) by homologous recombination (thick arrows) and between organisms of different populations by diverse horizontal transfer mechanisms (thin arrows). Genes that are only occasionally required in the metacommunity's environment may survive in the long term via transient associations with different genomes.

ecotypes carrying different GIs have been well established in the marine phototroph *Prochlorococcus*, with the different ecotypes coexisting at different frequencies along diverse environmental gradients. However, GIs have been shown to be highly variable in gene content among co-occurring cells of the same *Prochlorococcus* ecotype, suggesting that there is no strict correspondence between accessory gene content and niche differentiation (Coleman *et al.*, 2006). Genomic diversity at the population level has been analysed in detail in a unique study of the relatively simple microbial community that lives in an acid mine drainage system (Simmons *et al.*, 2008). In this case, the populations of the dominant species, *Ferroplasma* and *Leptospirillum*, consist of various genomic types that differ by numerous small insertions and blocs of divergent sequence, but population genetics tests indicate that the different variants have not diverged under positive selection and probably do not represent niche-differentiated ecotypes. In fact, the population genomic structure is consistent with the existence of a few abundant types that undergo homologous recombination plus a majority of low frequency variants. The patterns of genomic variation in *Prochlorococcus, Ferroplasma* and *Leptospirillum* do not fit the classical stable ecotype model, under which genomic diversity within coexisting ecotypes adapted to differentiated niches should be purged by selective sweeps and recombination among them should be limited (Cohan, 2006). Rather, they suggest that many accessory genes might indeed represent transient genome residents that do not reach fixation within local populations. Extensive temporal and geographic sampling of natural microbial communities at variable scales may be able to reveal whether specific accessory genes are stably or preferentially associated with certain genomic variants, or whether their distribution suggests a more nomadic mode of survival, dependent on frequent transfer among variants and/or repeated acquisition via immigrants from different communities (Fig. 1.1).

The fate of nomadic accessory genes should be linked to the interaction among their effects on the fitness of different host genomes, the variability of the environment and their transmissibility within and among species boundaries. For instance, accessory genes that provide strong fitness advantages under transient conditions may temporarily reach high frequencies within local populations or communities, and their long-term maintenance may be ensured by their capacity to disperse to other microbial communities during such periods of high abundance. This may be the case of accessory genes that confer resistance to broad range antibiotics or other strong selective agents that affect a variety of species but have a patchy distribution. Accessory genes whose presence is neutral or detrimental relative to organismal fitness may be able to survive via high levels of transmissibility, conferred by the different conditions capable of enhancing duplicability and transferability that have been described above. In the extreme, selfish, highly transmissible parasitic genes could maintain themselves by constructing their own niches within the genome, as has been proposed for different types of addiction modules (Naito *et al.*, 1995; Kobayashi, 2001). Mathematical modelling will be required to investigate the combinations of fitness effects on host genomes, environmental variability and gene transmissibility under which the survival of different types of accessory genes can be ensured.

The emergent genome

Within the described context of semiautonomous genes struggling for survival, the genome appears as an emergent entity, a moving picture of gene associations, some of which will endure, while others will be transient in nature. This malleability does not imply that genomes are simply bags of genes; on the contrary, it is at the genome level that the interactions between different genetic units are encoded so that they can function as a coherent whole. This picture is similar to that of ecological communities of organisms, which are capable of maintaining regulated dynamics and species interactions even if their exact species composition fluctuates through time and space. The capacity of genomes to engage in this type of flexibility may ensue from specific structural and organizational properties. From the structural point of view, variability among closely related genomes is often confined to very specific regions, while the general genomic framework is tightly conserved. Sequences and/or genomic architectures that serve as hotspots for recombination can provide a basis for such a pattern. On the other hand, gene interactions are embedded within complex networks whose topology may facilitate or hinder the incorporation or loss of accessory genes. Comparative analyses have shown that gene networks often contain a core of ancestral genes involved in large numbers of interactions (hubs) that are highly conserved across species, while genes that are progressively acquired during evolution encode less connected and less central proteins. Networks then seem to grow by acquiring genes in the periphery, often incoming through HGT (Pal *et al.*, 2005; Davids and Zhang, 2008; Kreimer *et al.*, 2008; D'Antonio and Ciccarelli, 2011). This network topology and mode of growth clearly enable a flux of accessory genes onto a core genomic framework, allowing for niche exploration and adaptation to changing environments.

Another common property of gene networks that facilitates the exchange of accessory components is modularity. Bacterial regulatory and metabolic networks are often organized in well-defined modules, sets of genes or proteins that are strongly interconnected and whose function is separable from those of other modules. This property is believed to be one of the main contributors to the robustness and evolvability of biological networks (Hintze and Adami, 2008). In fact, it has been shown that the extent of network modularity correlates with the breadth of the niche occupied by a bacterial species (Kreimer *et al.*, 2008). Simulation analyses have also shown that modularity allows for specialization in gene activity because it decreases interference between different groups of genes, and facilitates co-option, the utilization of existing gene activity to build new functional patterns (Espinosa-Soto and Wagner, 2010).

Clearly, genomes are shaped by structural and organizational properties that ensure their existence as coherent levels of organization in the face of the malleability conferred by gene duplicability and horizontal transfer. Such properties should evolve in response to natural selection at the organismal level, as they regulate gene interactions and permanence within genomes. Because architectural compartmentalization, network modularity and other genome-level properties enable the organized gain, loss and reshaping of activities that influence organismal survival, selection should favour their appearance, maintenance and refinement, especially under conditions of frequent environmental change. Furthermore, the exchange of genes among individuals belonging to the same population, community or metacommunity (Fig. 1.1) generates genetic relatedness within these different entities and diversifies the levels of organization at which natural selection may act. While gene flow hinders genetic and ecological divergence within populations, interspecific gene exchange enables the existence of a supraspecific gene pool that affects the likelihood of survival of individual organisms from different populations cohabiting the same ecological community or linked within a metacommunity. The constituency of this supraspecific gene pool will be shaped by the taxonomic composition of the ecological community, the specific genes present in the community metagenome at a given time and the rate of immigration of organisms and phages from other communities, while the MGEs present in the metacommunity will shape the type, rate and directionality of the HGT processes that distribute the gene pool. Consequently, organisms from different species coexisting in the same community or metacommunity may affect each other's fitness, not only through their ecological interactions, but also through their contribution to the common gene pool. In the long term, the survival of entire communities and metacommunities may depend on their metagenomic pool of transferable, accessory genes and on the type and rate of HGT processes that distribute it among organisms facing similar selective pressures.

Acknowledgement

I would like to thank Yvonne Vallès for her help with the references cited in this chapter.

References

Alcaraz, L.D., Olmedo, G., Bonilla, G., Cerritos, R., Hernandez, G., Cruz, A., Ramirez, E., Putonti, C., Jimenez, B., Martinez, E., *et al.* (2008). The genome of *Bacillus coahuilensis* reveals adaptations essential for survival in the relic of an ancient marine environment. Proc. Natl. Acad. Sci. U.S.A. *105*, 5803–5808.

Anderson, R.P., and Roth, J.R. (1977). Tandem genetic duplications in phage and bacteria. Annu. Rev. Microbiol. *31*, 473–505.

Avrani, S., Wurtzel, O., Sharon, I., Sorek, R., and Lindell, D. (2011). Genomic island variability facilitates *Prochlorococcus*-virus coexistence. Nature *474*, 604–608.

Berglund, E.C., Frank, A.C., Calteau, A., Vinnere Pettersson, O., Granberg, F., Eriksson, A.S., Naslund, K., Holmberg, M., Lindroos, H., and Andersson, S.G. (2009). Run-off replication of host-adaptability genes is associated with gene transfer agents in the genome of mouse-infecting *Bartonella grahamii*. PLoS Genet. *5*, e1000546.

Besenmatter, W., Kast, P., and Hilvert, D. (2007). Relative tolerance of mesostable and thermostable protein homologs to extensive mutation. Proteins *66*, 500–506.

Bloom, J.D., Labthavikul, S.T., Otey, C.R., and Arnold, F.H. (2006). Protein stability promotes evolvability. Proc. Natl. Acad. Sci. U.S.A. *103*, 5869–5874.

Boyd, E.F., and Hartl, D.L. (1998). *Salmonella* virulence plasmid. Modular acquisition of the *spv* virulence region by an F-plasmid in *Salmonella enterica* subspecies I and insertion into the chromosome of subspecies II, IIIa, IV and VII isolates. Genetics *149*, 1183–1190.

Boyd, E.F., Nelson, K., Wang, F.S., Whittam, T.S., and Selander, R.K. (1994). Molecular genetic basis of allelic polymorphism in malate dehydrogenase (mdh) in natural populations of *Escherichia coli* and *Salmonella enterica*. Proc. Natl. Acad. Sci. U.S.A. *91*, 1280–1284.

Brochier, C., Bapteste, E., Moreira, D., and Philippe, H. (2002). Eubacterial phylogeny based on translational apparatus proteins. Trends Genet. *18*, 1–5.

Brown, J.R., Douady, C.J., Italia, M.J., Marshall, W.E., and Stanhope, M.J. (2011). Universal trees based on large combined protein sequence data sets. Nat. Genet. *28*, 281-285.

Charlebois, R.L., and Doolittle, W.F. (2004). Computing prokaryotic gene ubiquity: rescuing the core from extinction. Genome Res. *14*, 2469–2477.

Chistoserdova, L., Vorholt, J.A., Thauer, R.K., and Lidstrom, M.E. (1998). C1 transfer enzymes and coenzymes linking methylotrophic bacteria and methanogenic Archaea. Science *281*, 99–102.

Ciccarelli, F.D., Doerks, T., von Mering, C., Creevey, C.J., Snel, B., and Bork, P. (2006). Toward automatic reconstruction of a highly resolved tree of life. Science *311*, 1283–1287.

Cohan, F.M. (2001). Bacterial species and speciation. Syst. Biol. *50*, 513–524.

Cohan, F.M. (2006). Towards a conceptual and operational union of bacterial systematics, ecology, and evolution. Phil. Trans. R Soc. Lond. B Biol. Sci. *361*, 1985–1996.

Coleman, M.L., Sullivan, M.B., Martiny, A.C., Steglich, C., Barry, K., Delong, E.F., and Chisholm, S.W. (2006). Genomic islands and the ecology and evolution of *Prochlorococcus*. Science *311*, 1768–1770.

Conant, G.C., and Wagner, A. (2002). GenomeHistory: a software tool and its application to fully sequenced genomes. Nucleic Acids Res. *30*, 3378–3386.

Cordero, O.X., and Hogeweg, P. (2009). The consequences of base pair composition biases for regulatory network organization in prokaryotes. Mol. Biol. Evol. *26*, 2171–2173.

de la Cruz, F., and Davies, J. (2000). Horizontal gene transfer and the origin of species: lessons from bacteria. Trends Microbiol. *8*, 128–133.

D'Antonio, M., and Ciccarelli, F.D. (2011). Modification of gene duplicability during the evolution of protein interaction network. PLoS Comput. Biol. *7*, e1002029.

Daubin, V., and Ochman, H. (2004). Bacterial genomes as new gene homes: the genealogy of ORFans in *E. Coli*. Genome Res. *14*, 1036-1042.

Davids, W., and Zhang, Z. (2008). The impact of horizontal gene transfer in shaping operons and protein interaction networks – direct evidence of preferential attachment. BMC Evol. Biol. *8*, 23.

Davis, J.C., and Petrov, D.A. (2004). Preferential duplication of conserved proteins in eukaryotic genomes. PLoS Biol. *2*, E55.

Dunham, M.J., Badrane, H., Ferea, T., Adams, J., Brown, P.O., Rosenzweig, F., and Botstein, D. (2002). Characteristic genome rearrangements in experimental evolution of *Saccharomyces cerevisiae*. Proc. Natl. Acad. Sci. U.S.A. *99*, 16144-16149.

Dunn, K.A., Bielawski, J.P., Ward, T.J., Urquhart, C., and Gu, H. (2009). Reconciling ecological and genomic divergence among lineages of *Listeria* under an 'extended mosaic genome concept'. Mol. Biol. Evol. *26*, 2605–2615.

Eberhard, W.G. (1990). Evolution in bacterial plasmids and levels of selection. Q. Rev. Biol. *65*, 3–22.

Espinosa-Soto, C., and Wagner, A. (2010). Specialization can drive the evolution of modularity. PLoS Comput. Biol. *6*, e1000719.

Fang, G., Rocha, E.P., and Danchin, A. (2008). Persistence drives gene clustering in bacterial genomes. BMC Genomics *9*, 4.

Feavers, I.M., Heath, A.B., Bygraves, J.A., and Maiden, M.C. (1992). Role of horizontal genetic exchange in the antigenic variation of the class 1 outer membrane protein of *Neisseria meningitidis*. Mol. Microbiol. *6*, 489–495.

Filloux, A. (2010). A variety of bacterial pili involved in horizontal gene transfer. J. Bacteriol. *192*, 3243–3245.

Francino, M.P. (2005). An adaptive radiation model for the origin of new gene functions. Nat. Genet. *37*, 573–577.

Froula, J.L., and Francino, M.P. (2007). Selection against spurious promoter motifs correlates with translational efficiency across bacteria. PLoS One *2*, e745.

Gerstein, M. (1997). A structural census of genomes: comparing bacterial, eukaryotic, and archaeal genomes in terms of protein structure. J. Mol. Biol. *274*, 562–576.

Gevers, D., Vandepoele, K., Simillon, C., and Van de Peer, Y. (2004). Gene duplication and biased functional retention of paralogs in bacterial genomes. Trends Microbiol. *12*, 148–154.

Gogarten, J.P., Doolittle, W.F., and Lawrence, J.G. (2002). Prokaryotic evolution in light of gene transfer. Mol. Biol. Evol. *19*, 2226–2238.

Gomez-Valero, L., Rocha, E.P., Latorre, A., and Silva, F.J. (2007). Reconstructing the ancestor of *Mycobacterium leprae*: the dynamics of gene loss and genome reduction. Genome Res. *17*, 1178–1185.

Hacker, J., and Kaper, J.B. (2000). Pathogenicity islands and the evolution of microbes. Annu. Rev. Microbiol. *54*, 641–679.

Haiko, J., Laakkonen, L., Westerlund-Wikstrom, B., and Korhonen, T.K. (2011). Molecular adaptation of a plant-bacterium outer membrane protease towards plague virulence factor Pla. BMC Evol. Biol. *11*, 43.

Hall, B.G., and Sharp, P.M. (1992). Molecular population genetics of *Escherichia coli*: DNA sequence diversity at the *celC*, *crr*, and *gutB* loci of natural isolates. Mol. Biol. Evol. *9*, 654–665.

Hansen, J.K., and Forest, K.T. (2006). Type IV pilin structures: insights on shared architecture, fiber assembly, receptor binding and type II secretion. J. Mol. Microbiol. Biotechnol. *11*, 192–207.

Harbaugh, M.P., Podbielski, A., Hugl, S., and Cleary, P.P. (1993). Nucleotide substitutions and small-scale insertion produce size and antigenic variation in group A streptococcal M1 protein. Mol. Microbiol. *8*, 981–991.

Harrison, P.M., and Gerstein, M. (2002). Studying genomes through the aeons: protein families, pseudogenes and proteome evolution. J. Mol. Biol. *318*, 1155–1174.

Hayashi, T., Makino, K., Ohnishi, M., Kurokawa, K., Ishii, K., Yokoyama, K., Itan, C.G., Ohtsubo, E., Nakayama, K., Murata, T., *et al.* (2001). Complete genome sequence of enterohemorrhagic *Escherichia coli* 0157OH7 and genomic comparison with a laboratory strain K-12. DNA Res. *8*, 11-22.

He, X., and Zhang, J. (2006). Higher duplicability of less important genes in yeast genomes. Mol. Biol. Evol. *23*, 144–151.

Hintze, A., and Adami, C. (2008). Evolution of complex modular biological networks. PLoS Comput. Biol. *4*, e23.

Hooper, S.D., and Berg, O.G. (2003a). Duplication is more common among laterally transferred genes than among indigenous genes. Genome Biol. *4*, R48.

Hooper, S.D., and Berg, O.G. (2003b). On the nature of gene innovation: duplication patterns in microbial genomes. Mol. Biol. Evol. *20*, 945–954.

Huerta, A.M., Francino, M.P., Morett, E., and Collado-Vides, J. (2006). Selection for unequal densities of sigma70 promoter-like signals in different regions of large bacterial genomes. PLoS Genet. *2*, e185.

Huynen, M.A., and van Nimwegen, E. (1998). The frequency distribution of gene family sizes in complete genomes. Mol. Biol. Evol. *15*, 583–589.

Ikemura, T. (1985). Codon usage and tRNA content in unicellular and multicellular organisms. Mol. Biol. Evol. *2*, 13–34.

Jacobsen, A., Hendriksen, R.S., Aaresturp, F.M., Ussery, D.W., and Friis, C. (2011). The *Salmonella enterica* pan-genome. Microb. Ecol. *62*, 487–504.

Jain, R., Rivera, M.C., and Lake, J.A. (1999). Horizontal gene transfer among genomes: the complexity hypothesis. Proc. Natl. Acad. Sci. U.S.A. *96*, 3801–3806.

Jordan, I.K., Makarova, K.S., Spouge, J.L., Wolf, Y.I., and Koonin, E.V. (2001). Lineage-specific gene expansions in bacterial and archaeal genomes. Genome Res. *11*, 555–565.

Jordan, I.K., Wolf, Y.I., and Koonin, E.V. (2004). Duplicated genes evolve slower than singletons despite the initial rate increase. BMC Evol. Biol. *4*, 22.

Kanaya, S., Yamada, Y., Kudo, Y., and Ikemura, T. (1999). Studies of codon usage and tRNA genes of 18 unicellular organisms and quantification of *Bacillus subtilis* tRNAs: gene expression level and species-specific diversity of codon usage based on multivariate analysis. Gene *238*, 143–155.

Karev, G.P., Wolf, Y.I., and Koonin, E.V. (2003). Simple stochastic birth and death models of genome evolution: was there enough time for us to evolve? Bioinformatics *19*, 1889–1900.

Kobayashi, I. (2001). Behavior of restriction-modification systems as selfish mobile elements and their impact on genome evolution. Nucleic Acids Res. *29*, 3742–3756.

Kondrashov, F.A., Rogozin, I.B., Wolf, Y.I., and Koonin, E.V. (2002). Selection in the evolution of gene duplications. Genome Biol. *3*, RESEARCH0008.

Kreimer, A., Borenstein, E., Gophna, U., and Ruppin, E. (2008). The evolution of modularity in bacterial metabolic networks. Proc. Natl. Acad. Sci. U.S.A. *105*, 6976–6981.

Lawrence, J.G. (1997). Selfish operons and speciation by gene transfer. Trends Microbiol. *5*, 355–359.

Lawrence, J.G. (1999). Gene transfer, speciation, and the evolution of bacterial genomes. Curr. Opin. Microbiol. *2*, 519–523.

Lawrence, J.G. (2002). Gene transfer in bacteria: speciation without species? Theor. Popul. Biol. *61*, 449–460.

Lawrence, J.G., and Roth, J.R. (1996). Selfish operons: horizontal transfer may drive the evolution of gene clusters. Genetics *143*, 1843–1860.

Lenski, R.E., and May, R.M. (1994). The evolution of virulence in parasites and pathogens: reconciliation between two competing hypotheses. J. Theor. Biol. *169*, 253–265.

Lerat, E., Daubin, V., and Moran, N.A. (2003). From gene trees to organismal phylogeny in prokaryotes: the case of the gamma-Proteobacteria. PLoS Biol. *1*, E19.

Lercher, M.J., and Pal, C. (2008). Integration of horizontally transferred genes into regulatory interaction networks takes many million years. Mol. Biol. Evol. *25*, 559–567.

Lespinet, O., Wolf, Y.I., Koonin, E.V., and Aravind, L. (2002). The role of lineage-specific gene family expansion in the evolution of eukaryotes. Genome Res. *12*, 1048–1059.

Levin, B.R., and Cornejo, O.E. (2009). The population and evolutionary dynamics of homologous gene recombination in bacterial populations. PLoS Genet. *5*, e1000601.

Li, J., Nelson, K., McWhorter, A.C., Whittam, T.S., and Selander, R.K. (1994). Recombinational basis of serovar diversity in *Salmonella enterica*. Proc. Natl. Acad. Sci. U.S.A. *91*, 2552–2556.

Li, L., Huang, Y., Xia, X., and Sun, Z. (2006). Preferential duplication in the sparse part of yeast protein interaction network. Mol. Biol. Evol. *23*, 2467-2473.

Li, Y., Youngren, B., Sergueev, K., and Austin, S. (2003). Segregation of the *Escherichia coli* chromosome terminus. Mol. Microbiol. *50*, 825-834.

Lin, Y.S., Hwang, J.K., and Li, W.H. (2007). Protein complexity, gene duplicability and gene dispensability in the yeast genome. Gene *387*, 109–117.

Lind, P.A., Tobin, C., Berg, O.G., Kurland, C.G., and Andersson, D.I. (2010). Compensatory gene amplification restores fitness after inter-species gene replacements. Mol. Microbiol. *75*, 1078–1089.

Livermore, D.M. (1995). Beta-Lactamases in laboratory and clinical resistence. Clin. Mircobial. Rev. *8*, 557-584.

Lucchini, S., Rowley, G., Goldberg, M.D., Hurd, D., Harrison, M., and Hinton, J.C. (2006). H-NS mediates the silencing of laterally acquired genes in bacteria. PLoS Pathog. *2*, e81.

Lukjancenko, O., Wassenaar, T.M., and Ussery, D.W. (2010). Comparison of 61 sequenced *Escherichia coli* genomes. Microb. Ecol. *60*, 708–720.

Ma, W., Dong, F.F., Stavrinides, J., and Guttman, D.S. (2006). Type III effector diversification via both pathoadaptation and horizontal transfer in response to a coevolutionary arms race. PLoS Genet. *2*, e209.

Makarova, K.S., Wolf, Y.I., and Koonin, E.V. (2009). Comprehensive comparative-genomic analysis of type 2 toxin–antitoxin systems and related mobile stress response systems in prokaryotes. Biol. Direct *4*, 19.

Manna, D., Breier, A.M., and Higgins, N.P. (2004). Microarray analysis of transposition targets in *Escherichia coli*: the impact of transcription. Proc. Natl. Acad. Sci. U.S.A. *101*, 9780–9785.

Marchetti, M., Capela, D., Glew, M., Cruveiller, S., Chane-Woon-Ming, B., Gris, C., Timmers, T., Poinsot, V., Gilbert, L.B., Heeb, P., *et al.* (2010). Experimental evolution of a plant pathogen into a legume symbiont. PLoS Biol. *8*, e1000280.

Marklund, B.I., Tennent, J.M., Garcia, E., Hamers, A., Baga, M., Lindberg, F., Gaastra, W., and Normark, S. (1992). Horizontal gene transfer of the *Escherichia coli pap* and *prs* pili operons as a mechanism for the development of tissue-specific adhesive properties. Mol. Microbiol. *6*, 2225–2242.

Marland, E., Prachumwat, A., Maltsev, N., Gu, Z., and Li, W.H. (2004). Higher gene duplicabilities for metabolic proteins than for nonmetabolic proteins in yeast and *E. coli*. J. Mol. Evol. *59*, 806–814.

Mira, A., Ochman, H., and Moran, N.A. (2001). Deletional bias and the evolution of bacterial genomes. Trends Genet. *17*, 589–596.

Mira, A., Martin-Cuadrado, A.B., D'Auria, G., and Rodriguez-Valera, F. (2010). The bacterial pan-genome: a new paradigm in microbiology. Int. Microbiol. *13*, 45–57.

Mochizuki, A., Yahara, K., Kobayashi, I., and Iwasa, Y. (2006). Genetic addiction: selfish gene's strategy for symbiosis in the genome. Genetics *172*, 1309–1323.

Moran, N.A. (1996). Accelerated evolution and Muller's rachet in endosymbiotic bacteria. Proc. Natl. Acad. Sci. U.S.A. *93*, 2873–2878.

Murray, N.E., Daniel, A.S., Cowan, G.M., and Sharp, P.M. (1993). Conservation of motifs within the unusually variable polypeptide sequences of type I restriction and modification enzymes. Mol. Microbiol. *9*, 133–143.

Naito, T., Kusano, K., and Kobayashi, I. (1995). Selfish behavior of restriction–modification systems. Science *267*, 897–899.

Nakamura, Y., Itoh, T., Matsuda, H., and Gojobori, T. (2004). Biased biological functions of horizontally transferred genes in prokaryotic genomes. Nat. Genet. *36*, 760–766.

Navarre, W.W., Porwollik, S., Wang, Y., McClelland, M., Rosen, H., Libby, S.J., and Fang, F.C. (2006). Selective silencing of foreign DNA with low GC content by the H-NS protein in *Salmonella*. Science *313*, 236–238.

Navarre, W.W., McClelland, M., Libby, S.J., and Fang, F.C. (2007). Silencing of xenogeneic DNA by H-NS-facilitation of lateral gene transfer in bacteria by a defense system that recognizes foreign DNA. Genes Dev. *21*, 1456–1471.

Nogueira, T., Rankin, D.J., Touchon, M., Taddei, F., Brown, S.P., and Rocha, E.P. (2009). Horizontal gene transfer of the secretome drives the evolution of bacterial cooperation and virulence. Curr. Biol. *19*, 1683–1691.

Normand, P., Lapierre, P., Tisa, L.S., Gogarten, J.P., Alloisio, N., Bagnarol, E., Bassi, C.A., Berry, A.M., Bickhart, D.M., Choisne, N., *et al.* (2007). Genome characteristics of facultatively symbiotic *Frankia* sp. strains reflect host range and host plant biogeography. Genome Res. *17*, 7–15.

Ochman, H., and Groisman, E.A. (1996). Distribution of pathogenicity islands in *Salmonella* spp. Infect. Immun. *64*, 5410–5412.

Ochman, H., Lawrence, J.G., and Groisman, E.A. (2000). Lateral gene transfer and the nature of bacterial innovation. Nature *405*, 299–304.

Ogier, J.C., Calteau, A., Forst, S., Goodrich-Blair, H., Roche, D., Rouy, Z., Suen, G., Zumbihl, R., Givaudan, A., Tailliez, P., *et al.* (2010). Units of plasticity in bacterial genomes: new insight from the comparative genomics of two bacteria interacting with invertebrates, *Photorhabdus* and *Xenorhabdus*. BMC Genomics *11*, 568.

Omelchenko, M.V., Makarova, K.S., Wolf, Y.I., Rogozin, I.B., and Koonin, E.V. (2003). Evolution of mosaic operons by horizontal gene transfer and gene displacement *in situ*. Genome Biol. *4*, R55.

Omer, S., Kovacs, A., Mazor, Y., and Gophna, U. (2010). Integration of a foreign gene into a native complex does not impair fitness in an experimental model of lateral gene transfer. Mol. Biol. Evol. *27*, 2441–2445.

Pal, C., and Hurst, L.D. (2004). Evidence against the selfish operon theory. Trends Genet. *20*, 232–234.

Pal, C., Papp, B., and Lercher, M.J. (2005). Adaptive evolution of bacterial metabolic networks by horizontal gene transfer. Nat. Genet. *37*, 1372–1375.

Papp, B., Pal, C., and Hurst, L.D. (2003). Dosage sensitivity and the evolution of gene families in yeast. Nature *424*, 194–197.

Patil, P.B., Bogdanove, A.J., and Sonti, R.V. (2007). The role of horizontal transfer in the evolution of a highly variable lipopolysaccharide biosynthesis locus in xanthomonads that infect rice, citrus and crucifers. BMC Evol. Biol. *7*, 243.

Pelicic, V. (2008). Type IV pili: e pluribus unum? Mol. Microbiol. *68*, 827–837.

Penn, K., Jenkins, C., Nett, M., Udwary, D.W., Gontang, E.A., McGlinchey, R.P., Foster, B., Lapidus, A., Podell, S., Allen, E.E., *et al.* (2009). Genomic islands link secondary metabolism to functional adaptation in marine Actinobacteria. ISME J *3*, 1193–1203.

Perna, N.T., Plunkett, G., 3rd, Burland, V., Mau, B., Glasner, J.D., Rose, D.J., Mayhew, G.F., Evans, P.S., Gregor, J., Kirkpatrick, H.A., *et al.* (2001). Genome sequence of enterohaemorrhagic *Escherichia coli* O157:H7. Nature *409*, 529–533.

Prachumwat, A., and Li, W.H. (2006). Protein function, connectivity, and duplicability in yeast. Mol. Biol. Evol. *23*, 30–39.

Price, M.N., Huang, K.H., Arkin, A.P., and Alm, E.J. (2005). Operon formation is driven by co-regulation and not by horizontal gene transfer. Genome Res. *15*, 809–819.

Qian, J., Luscombe, N.M., and Gerstein, M. (2001). Protein family and fold occurrence in genomes: power-law behaviour and evolutionary model. J. Mol. Biol. *313*, 673–681.

Rankin, D.J., Rocha, E.P., and Brown, S.P. (2011). What traits are carried on mobile genetic elements, and why? Heredity (Edinb.) *106*, 1–10.

Reams, A.B., and Neidle, E.L. (2003). Genome plasticity in *Acinetobacter*: new degradative capabilities acquired by the spontaneous amplification of large chromosomal segments. Mol. Microbiol. *47*, 1291–1304.

Reams, A.B., Kofoid, E., Savageau, M., and Roth, J.R. (2010). Duplication frequency in a population of *Salmonella enterica* rapidly approaches steady state with or without recombination. Genetics *184*, 1077–1094.

Riehle, M.M., Bennett, A.F., and Long, A.D. (2001). Genetic architecture of thermal adaptation in *Escherichia coli*. Proc. Natl. Acad. Sci. U.S.A. *98*, 525–530.

Riley, M.A., and Wertz, J.E. (2002). Bacteriocin diversity: ecological and evolutionary perspectives. Biochimie *84*, 357–364.

Rocha, E.P. (2004). Codon usage bias from tRNA's point of view: redundancy, specialization, and efficient decoding for translation optimization. Genome Res. *14*, 2279–2286.

Rocha, E.P., and Danchin, A. (2002). Base composition bias might result from competition for metabolic resources. Trends Genet. *18*, 291–294.

Romero, D., and Palacios, R. (1997). Gene amplification and genomic plasticity in prokaryotes. Annu. Rev. Genet. *31*, 91–111.

Saavedra De Bast, M., Mine, N., and Van Melderen, L. (2008). Chromosomal toxin–antitoxin systems may act as antiaddiction modules. J. Bacteriol. *190*, 4603–4609.

Santos, S.R., and Ochman, H. (2004). Identification and phylogenetic sorting of bacterial lineages with universally conserved genes and proteins. Environ. Microbiol. *6*, 754–759.

Schmidt, M.A. (2010). LEEways: tales of EPEC, ATEC and EHEC. Cell Microbiol. *12*, 1544–1552.

Sharp, P.M., and Li, W.H. (1986). An evolutionary perspective on synonymous codon usage in unicellular organisms. J. Mol. Evol. *24*, 28–38.

Silby, M.W., Winstanley, C., Godfrey, S.A., Levy, S.B., and Jackson, R.W. (2011). *Pseudomonas* genomes: diverse and adaptable. FEMS Microbiol. Rev. *35*, 652–680.

Simmons, S.L., Dibartolo, G., Denef, V.J., Goltsman, D.S., Thelen, M.P., and Banfield, J.F. (2008). Population genomic analysis of strain variation in *Leptospirillum* group II bacteria involved in acid mine drainage formation. PLoS Biol. *6*, e177.

Smith, N.H., and Selander, R.K. (1991). Molecular genetic basis for complex flagellar antigen expression in a triphasic serovar of *Salmonella*. Proc. Natl. Acad. Sci. U.S.A. *88*, 956–960.

Sobecky, P.A., and Coombs, J.M. (2009). Horizontal gene transfer in metal and radionuclide contaminated soils. Methods Mol. Biol. *532*, 455–472.

Sorek, R., Zhu, Y., Creevey, C.J., Francino, M.P., Bork, P., and Rubin, E.M. (2007). Genome-wide experimental determination of barriers to horizontal gene transfer. Science *318*, 1449–1452.

Spratt, B.G., Bowler, L.D., Zhang, Q.Y., Zhou, J., and Smith, J.M. (1992). Role of interspecies transfer of chromosomal genes in the evolution of penicillin resistance in pathogenic and commensal *Neisseria* species. J. Mol. Evol. *34*, 115–125.

Stewart, F.M., and Levin, B.R. (1977). The population biology of bacterial plasmids: a PRIORI conditions for the existence of conjugationally transmitted factors. Genetics *87*, 209–228.

Stinear, T.P., Seemann, T., Pidot, S., Frigui, W., Reysset, G., Garnier, T., Meurice, G., Simon, D., Bouchier, C., Ma, L., *et al.* (2007). Reductive evolution and niche adaptation inferred from the genome of *Mycobacterium* ulcerans, the causative agent of Buruli ulcer. Genome Res. *17*, 192–200.

Stinear, T.P., Seemann, T., Harrison, P.F., Jenkin, G.A., Davies, J.K., Johnson, P.D., Abdellah, Z., Arrowsmith, C., Chillingworth, T., Churcher, C., *et al.* (2008). Insights from the complete genome sequence of *Mycobacterium marinum* on the evolution of *Mycobacterium tuberculosis*. Genome Res. *18*, 729–741.

Tettelin, H., Masignani, V., Cieslewicz, M.J., Donati, C., Medini, D., Ward, N.L., Angiuoli, S.V., Crabtree, J., Jones, A.L., Durkin, A.S., *et al.* (2005). Genome analysis of multiple pathogenic isolates of *Streptococcus agalactiae*: implications for the microbial 'pan-genome'. Proc. Natl. Acad. Sci. U.S.A. *102*, 13950–13955.

Thompson, J.R., Pacocha, S., Pharino, C., Klepac-Ceraj, V., Hunt, D.E., Benoit, J., Sarma-Rupavtarm, R., Distel, D.L., and Polz, M.F. (2005). Genotypic diversity within a natural coastal bacterioplankton population. Science *307*, 1311–1313.

Tokuriki, N., and Tawfik, D.S. (2009). Protein dynamism and evolvability. Science *324*, 203–207.

Toth, I.K., Pritchard, L., and Birch, P.R. (2006). Comparative genomics reveals what makes an enterobacterial plant pathogen. Annu. Rev. Phytopathol. *44*, 305–336.

Touchon, M., Hoede, C., Tenaillon, O., Barbe, V., Baeriswyl, S., Bidet, P., Bingen, E., Bonacorsi, S., Bouchier, C., Bouvet, O., *et al.* (2009). Organised genome dynamics in the *Escherichia coli* species results in highly diverse adaptive paths. PLoS Genet. *5*, e1000344.

Tuller, T. (2011). Codon bias, tRNA pools and horizontal gene transfer. Mob. Genet. Elements *1*, 75–77.

Van Valen, L. (1974). Predation and species diversity. J. Theor. Biol. *44*, 19–21.

Vinuesa, P., Silva, C., Werner, D., and Martinez-Romero, E. (2005). Population genetics and phylogenetic inference in bacterial molecular systematics: the roles of migration and recombination in *Bradyrhizobium* species cohesion and delineation. Mol. Phylogenet. Evol. *34*, 29–54.

Wagner, A. (2005). Robustness, evolvability, and neutrality. FEBS Lett. *579*, 1772-1778.

Wall, D., and Kaiser, D. (1999). Type IV pili and cell motility. Mol. Microbiol. *32*, 1–10.

Wang, L., Rothemund, D., Curd, H., and Reeves, P.R. (2003). Species-wide variation in the *Escherichia coli* flagellin (H-antigen) gene. J. Bacteriol. *185*, 2936–2943.

Waterhouse, R.M., Zdobnov, E.M., and Kriventseva, E.V. (2011). Correlating traits of gene retention, sequence divergence, duplicability and essentiality in vertebrates, arthropods, and fungi. Genome Biol. Evol. *3*, 75–86.

Wellner, A., and Gophna, U. (2008). Neutrality of foreign complex subunits in an experimental model of lateral gene transfer. Mol. Biol. Evol. *25*, 1835–1840.

Wellner, A., Lurie, M.N., and Gophna, U. (2007). Complexity, connectivity, and duplicability as barriers to lateral gene transfer. Genome Biol. *8*, R156.

Wiedenbeck, J., and Cohan, F.M. (2011). Origins of bacterial diversity through horizontal genetic transfer and adaptation to new ecological niches. FEMS Microbiol. Rev. *35*, 957–976.

Wu, X., Monchy, S., Taghavi, S., Zhu, W., Ramos, J., and van der Lelie, D. (2011). Comparative genomics and functional analysis of niche-specific adaptation in *Pseudomonas putida*. FEMS Microbiol. Rev. *35*, 299–323.

Xiong, J., Inoue, K., and Bauer, C.E. (1998). Tracking molecular evolution of photosynthesis by characterization of a major photosynthesis gene cluster from *Heliobacillus mobilis*. Proc. Natl. Acad. Sci. U.S.A. *95*, 14851–14856.

Yadid, I., Kirshenbaum, N., Sharon, M., Dym, O., and Tawfik, D.S. (2010). Metamorphic proteins mediate evolutionary transitions of structure. Proc. Natl. Acad. Sci. U.S.A. *107*, 7287–7292.

Yang, J., Lusk, R., and Li, W.H. (2003). Organismal complexity, protein complexity, and gene duplicability. Proc. Natl. Acad. Sci. U.S.A. *100*, 15661–15665.

Zhong, S., Khodursky, A., Dykhuizen, D.E., and Dean, A.M. (2004). Evolutionary genomics of ecological specialization. Proc. Natl. Acad. Sci. U.S.A. *101*, 11719–11724.

Evolution of Prokaryotic Pangenomes

2

Florent Lassalle and Vincent Daubin

Abstract

Since Fox and Woese (1977) and the discovery of the two ancient 'kingdoms' of Bacteria and Archaea, molecular biologists have been constantly revising their perception of the abundance and variety of microbial life, and the processes driving their evolution. In recent years, genome sequences from closely related species and strains of the same species have revealed a yet unforeseen diversity of gene repertoires, much larger than could have been predicted by their morphologies or biochemical capabilities. The origins and the functional significance of these differences still need to be understood, and the techniques to address these questions have yet by large to be invented. In this chapter, we will present a review of the current knowledge on the mechanisms responsible for the diversity of gene repertoires in prokaryotes.

Gene repertoires and the organization of prokaryotic genomes

Prokaryotic genomes share global features that make them readily distinguishable from most eukaryotic ones. Eukaryotes are thought to accumulate DNA in their genomes by either duplication, invasion of transposable elements or insertion of endoviruses and usually harbour a high proportion of non-coding DNA. As a result, the coding capacity of a eukaryotic genome can be as low as 5%, for instance for the human genome. Prokaryotic genomes, on the other hand, are rather dense in genes, with usually around 80% of a genome coding for proteins. This compaction goes along with an optimized organization of the genetic information reflecting the action of cellular processes such as genome replication and gene expression. For instance, the asymmetric nature of DNA replication, with a lagging strand and a leading strand, impacts bacterial genome organization (Rocha and Danchin, 2003) as well as compositional and mutational patterns (Lobry and Sueoka, 2002). Similarly, the polarity of replication, from origin (*ori* locus) to terminus (*ter*), also leaves marks on circular bacterial chromosomes, as the composition in A+T nucleotides and evolutionary rates of genes near the terminus are generally higher (Daubin and Perrière, 2003). In the case of fast-growing bacteria, chromosomes replicate faster than the cell divides, which generates a strong selective pressure for relocating genes requiring high expression rates within the region of the origin of replication (Couturier and Rocha, 2006). Although the replication process in Archaea may differ in many points from the one observed in Bacteria, they share some of their characteristics, notably the strand composition asymmetry (Myllykallio *et al.*, 2000). Bacterial chromosomes also seem to be divided into regions differing in their average gene expression level. Functionally related genes are often co-transcribed in operons and sometimes further clustered in superoperons that can be widely conserved among prokaryotes, like ribosomal superoperons or pilus biosynthesis loci (Lathe *et al.*, 2000). Finally, chromosomal macro-domains, defined following the *ori-ter* axial symmetry, form units of megabase size that seem to be globally regulated through topological constraints (Esnault *et al.*, 2007).

As these features are conserved among the majority of prokaryotes, the constraints imposed by the replication and expression processes on the structure of bacterial chromosomes must be very strict. Indeed, disruption of the macro-domain symmetry appears to be highly detrimental in *E. coli* (Esnault *et al.*, 2007). The existence of strong selection for chromosomal organization in the whole bacterial kingdom is confirmed by the good conservation of general gene order between pairs of chromosomes of even moderately related strains (Rocha, 2006).

Despite these common organization principles, prokaryotic genomes are extremely diverse. Massive genome sequencing revealed an unexpected diversity of gene sequences. At the evolutionary scale of the whole prokaryotic world, hundreds of thousands of different genes have been inventoried in genomic databases (Penel *et al.*, 2009; Muller *et al.*, 2010). This number is immense, especially in regard to the estimation that fewer than 50 genes are conserved among all prokaryotes, most of which are involved in translation (Charlebois and Doolittle, 2004). All other processes, even those that are deemed essential, such as DNA replication and central metabolism, are apparently not homologous among all prokaryotes. This suggests that, through evolution, most of the ancestral genetic information has been replaced, at least in some lineages.

Although this pattern of loss and gain of genes may not be surprising at the scale of several billion years, the same applies at every phylogenetic depth, from phylum to species (Daubin and Ochman, 2004; Lerat *et al.*, 2005; van Passel *et al.*, 2008). Recent sequencing of several strains of the same species revealed that the gene content of strains within a species differs markedly while the overall genomic organization is conserved. The comparison of 20 strains of *E. coli* showed that the species' core genome – those genes common to all strains – consists of only ~2000 genes, less than half the average genome size of a strain (Touchon *et al.*, 2009). The remaining genes in each genome are shared by only a fraction of all strains, with the majority being unique to only one. This demonstrates that species are much more diverse than previously suggested by molecular phylogenies based on core genes, and raises the question of what makes the unity of species. The very notion of species is challenged by these facts and a definition for prokaryotic species is still sought and intensely debated (Achtman and Wagner, 2008). The extent of gene diversity within a species can simply be measured by looking at the size of its pangenome, which corresponds to all genes represented in at least one of the genomes of the considered species. For instance, the pangenome of *E. coli* is approximately 18,000 genes when considering currently sequenced genomes (Touchon *et al.*, 2009). Currently available pangenome data suggest that many species, such as *Streptococcus agalactiae* (Tettelin *et al., 2005), Vibrio cholerae* (Vesth *et al.*, 2010) or *Burkholderia pseudomallei* (Ussery *et al.*, 2009), have an 'open' pangenome where sampling of new genomes invariably reveals the existence of new genes. In contrast, other species seem to have limited gene repertoire diversity, such as *Bacillus anthracis*, for which no new gene has been found after sampling of four strains, indicating a 'closed' pangenome. Similar results were obtained in another recent study using data from almost one hundred genomes of the two sister species *Campylobacter jejuni* and *C. coli,* which established their common pangenome as 'closed' at ~3000 genes (Lefébure *et al.,* 2010). A closed pangenome usually means a recent speciation event with no genome diversification, but in the case of *Campylobacter,* it is unsure that the sequenced epidemic strains can be considered a good representative set of the existing species diversity (Lefébure *et al.,* 2010). The population pyramid of genes within a species is rather like a column, with pedestal and capital, i.e. peaks of frequency for genes restrained to a small set of genomes at the basis and for core genes at the top (Fig. 2.1) (Rocha, 2008; Touchon *et al.,* 2009; Lefébure *et al.,* 2010). This suggests that within a species, genomes are made of a relatively stable set of genes present in almost every genome (the 'stabilome') (Vesth *et al.,* 2010), and of a variable set of genes, which may even be strain-specific.

The diversity of bacterial pangenomes adds up to the immensity of the protein universe. In fact, the large set of homologous protein families obtained by merging all prokaryotic pangenomes displays a U-shaped distribution of frequencies similar to that obtained for genes among genomes (Koonin and Wolf, 2008; Lapierre and Gogarten, 2009). This shows that much of the protein diversity consists of sporadic proteins occurring in a very restricted

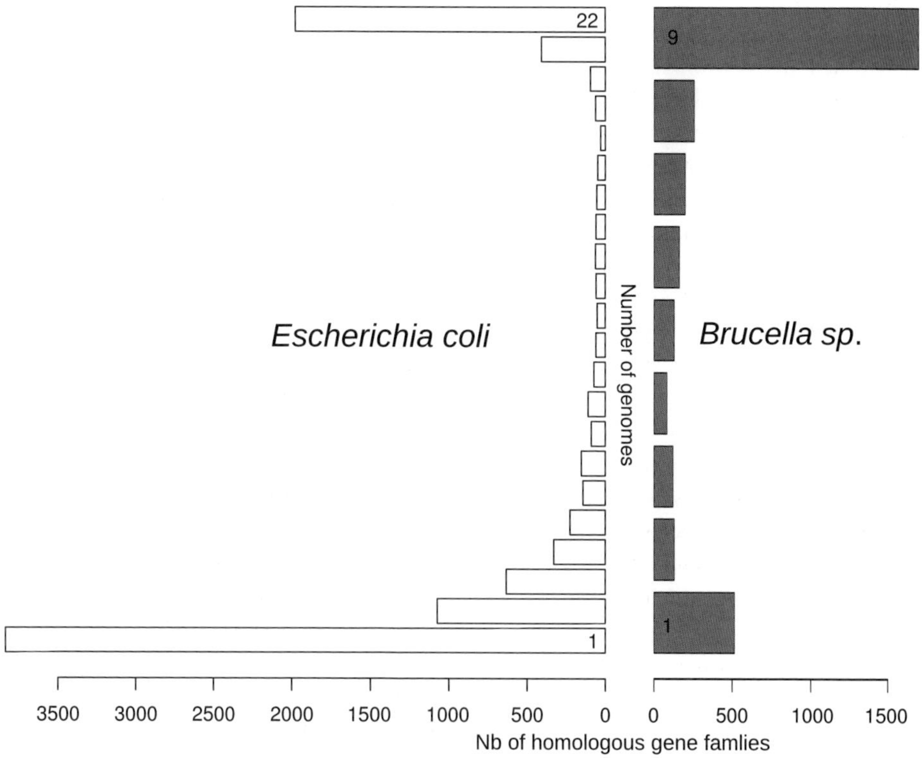

Figure 2.1 The pangenomes of *Escherichia coli* and *Brucella* sp. This plot represents the frequency of genes within 22 sequenced strains of *E. coli* and nine sequenced strains of *Brucella* sp., which includes genomes from five identified species (*B. melitentis*, *B. suis*, *B. abortus*, *B. ovis* and *B. canis*). These two pangenomes have very different dynamics, with a relatively small, closed pangenome in *Brucella* sp. and a large open one in *E. coli*. Data from Hogenom release 5 (http://pbil.univ-lyon1.fr/databases/hogenom/).

range of genomes, a large fraction of which correspond to ORFans, i.e. putative coding sequences which are unique to a genome.

How can bacterial genomes reconcile their optimized, constrained and highly conserved organization with so much variation in their content? Alignments of genomes of strains from the same species reveal they share a backbone of core genes with conserved order, inter-spaced with sporadic genes. Touchon and colleagues (2009) showed that in the *E. coli* species, most insertions occurred in a small set of discrete loci. This is in line with the assumption that the disruption of the genomic organization is highly detrimental, and only few sites allow neutral insertion of foreign DNA.

The dynamics of the pangenome

It has long been known that in many bacterial species large fractions of the genome have been acquired 'recently' by horizontal gene transfer (HGT) (Médigue *et al.*, 1991). These fractions are recognizable because the genes they contain show a codon usage which differs markedly from that of the rest of the genome. Such anomalous composition is thought to be a trace of the donor's genome composition, currently in 'amelioration' towards the normal codon use of the new host (Lawrence and Ochman, 1997). The variability of gene repertoires within a species and the elevated number of strain-specific genes suggest high rates of gene gain and loss within a lineage, typically higher than nucleotide substitution (Hao and Golding, 2006). For instance, when comparing two *E. coli* strains, orthologous sequences show less than 5% divergence at the nucleotide level, while gene content can differ by a thousand genes (~1 Mb) corresponding to more than 20% genomic dissimilarity (Touchon *et al.*, 2009). Gene gain and loss thus represent the major source of innovation in prokaryotic genomes, as the modifications of gene repertoires can strongly affect cellular metabolism and morphology. Such changes can efficiently provide prokaryotes with opportunities for adaptation to new ecological niches.

It is yet unclear how much of the observed differences among genomes are adaptive. The insertion of new genes in bacterial genomes is indeed thought to occur at very high frequency, either from duplication of resident genes or via HGT, the latter playing a dominant role (Snel *et al.*, 2002; Mirkin *et al.*, 2003; Kunin *et al.*, 2005; Treangen and Rocha, 2011). These mechanisms are thought to compensate the high deletion rate, which is universal in prokaryotes (Mira *et al.*, 2001; Kuo and Ochman, 2009), and to maintain a homogeneous genome size among strains within a species (Moran and Plague, 2004). Interestingly, obligatory intracellular pathogens or endosymbionts display spectacular reductions of their genomes (Ochman and Moran, 2001), reflecting both the absence of gene input by transfer and the increased effect of drift on slightly deleterious deletions (Ochman and Moran, 2001; Moran and Plague, 2004).

Once they enter the pangenome, new genes undergo rapid evolution. First, new genes are more prone to be deleted than older ones (Marri *et al.*, 2007; van Passel *et al.*, 2008). In addition, it has been shown that new genes have high evolutionary rates, and that this rate decreases with time spent in the genome (Ochman and Moran, 2001; Daubin and Ochman, 2004), which has been interpreted by Hao and Golding as 'a mixture of directional selection to adapt in some genes and neutral mutations destroying function in others' (Hao and Golding, 2006). Hence, newly acquired genes may not immediately be advantageous for the organism, and therefore be subject to relaxed selection. This can lead to either inactivation followed by rapid deletion or to fixation under a diverged form adapted to the new host environment. Even in free-living bacteria, many predicted open reading frames (ORFs) may be unsuspected decaying genes (Ochman and Davalos, 2006).

The roots of bacterial pangenomes

The consequence of this constant gene turnover is the cohabitation of genes of various origins and ages in a given lineage (Daubin and Ochman, 2004). Interestingly, within an age class, genes tend to be similar in terms of G+C%, length, and evolutionary rate (Daubin and Ochman, 2004; Hao and Golding, 2006; Yin and Fischer, 2008). ORFans, those genes that have no known homologues in any sequenced genome, i.e. the youngest genes, are short, fast evolving, and systematically AT-richer than their host genomes.

Three major hypothesis are presently discussed for the origin of ORFans: (1) origination *de novo* from non-coding sequences; (2) fast evolution of resident genes, to the point of losing any homology signal; (3) transfer from an inaccessible gene pool, consisting probably of mobile DNA elements such as viruses, plasmids, integrons or other integrative elements (IEs).

Origination *de novo* is not likely to happen in bacterial genomes, as stretches of non-coding DNA are short and mostly under selective constraints for their role in transcriptional regulation. Origin from fast evolution of pre-existing genes – resident or IEs – is difficult to test, as ORFans have no similarity to other sequences in databases. It has been suggested that ORFans have amino-acid compositions reminiscent of translation from random genomic sequences (Yomtovian *et al.*, 2010), but the nucleotide compositional bias inherent to ORFans may explain some of this resemblance. Recent studies by Eyre-Walker and Petrov show that mutations are universally biased towards AT nucleotides in Bacteria (Hershberg and Petrov, 2010; Hildebrand *et al.*, 2010). These authors concluded that the existence of GC-rich genes in genomes must therefore be explained by selection on GC content, or a selection-like process such as biased gene conversion (Hershberg and Petrov, 2010). Although a selective pressure acting on global genomic GC content is difficult to imagine, it could explain why ORFans, which are under lower purifying selection, have relatively low GC content.

There are a number of studies that suggest that ORFans come from IEs. IEs are known to be vectors for HGT by transduction/conjugation and frequently carry cellular genes from former hosts, and as such are also good candidates for transmitting new genes. ORFans have been shown to be frequently part of clusters of co-transferred genes that globally show similarities to IEs (Cortez *et al.*, 2009). The viral gene pool is today almost unexplored and first attempts to look into it suggest an incommensurable diversity (Edwards and Rohwer, 2005; Williamson *et al.*, 2008; Angly *et al.*, 2009) that could be an invaluable source of genes for prokaryotic genomes.

Functional role of the pangenome

The question remains of how many of these genes actually contribute to the fitness of their host. Those genes that are sporadically distributed among strains of a bacterial species are expected to have very marginal effects, and indeed genome streamlining experiments demonstrate that up to a forth of a bacterial genome can be deleted without affecting essential functions: by deleting genes that are poorly conserved among *E. coli* strains (Pósfai *et al.*, 2006), or genes absent in the related enterobacteria *Buchnera* (Mizoguchi *et al.*, 2008), *E. coli* strains can be engineered into fast-growing minimal cell factories (Mizoguchi *et al.*, 2007). Following a different approach, attempts to delete every single gene in genomes were conducted to assess viability and competitiveness of mutant cells (Gerdes *et al.*, 2003; Kobayashi *et al.*, 2003). In *E. coli* and *Bacillus subtilis,* respectively 620 and 271 genes were identified as essential, some of them being sporadic within the species (Gerdes *et al.*, 2003).

It thus appears that some newly acquired genes can persist and provide a substantial gain in fitness to their host. For instance, symbiotic genes that enable rhizobia to have a symbiotic relationship with plants are quite heterogeneously distributed among Proteobacteria (González *et al.*, 2003), but are fixed in numerous lineages that have adopted this highly beneficial lifestyle. This kind of genes with heterogeneous occurrence in prokaryotes (HOPs) have likely been transferred frequently while rarely settling stably in a species' genome. Indeed, it has been shown that genes providing advantages in specific environmental conditions in *E. coli* have a higher tendency to be transferred and lost across the tree of Proteobacteria (Pal *et al.*, 2005). The tendency of a gene to be transferred also correlates with the situation of its product in the metabolic (Pal *et al.*, 2005) or protein interactions networks (Davids and Zhang, 2008), with genes having few connections being more likely to be transferred. Such peripheral functions might mostly consist of interactions with the environment – like transport or sensing – and to be adaptive only under some specific ecological constraints. HOPs are likely retained when the environment provides sufficient selective advantage for such interactions, but will be lost if such conditions are only transient.

In certain cases, however, such genes accumulate in genomes and provide the cell with environmental versatility. This is the case in the group of the Rhizobiaceae, where large multireplicon genomes harbour the highest known diversity of ABC-type transporters, which allows these organisms to live on a high diversity of carbohydrate nutrients. In *Frankia,* host versatility has similarly been related to increased genome size (Normand *et al.*, 2007). In fact, it is possible that genes with high adaptive potential stay in pangenomes without reaching fixation, provided there exists a transient selective advantage, which insures episodic increase of their gene frequency. This would be particularly true if intraspecific recombination occurs frequently. Indeed, condition-specific or niche-specific genes seem to be the most mobile (Pal *et al.*, 2005), and sufficient transfer rates may prevent them from extinction by jumping from an organism to another. For instance, genes determining pathogenicity in *Vibrio cholerae* are associated with mobile elements like CTX phages and TCP pathogenicity islands (Rahman *et al.*, 2008). Pathogenicity in animals is only an accessory ecological niche of *V. cholerae,* which is essentially an aquatic microbe, but the determinants for pathogenicity remain constantly present in the environmental reservoir.

In some instances, transferred adaptive genes become fixed and transmitted vertically to all descendant taxa. The reconstruction of gene acquisition history on the lineage of *Salmonella enterica* sv. Typhimurium LT2 (Porwollik *et al.*, 2002) showed the gradual acquisition of pathogenicity determinants such as fimbriae, lipopolysaccharides or lipoprotein envelope biosynthesis genes during adaptation to the new warm-blood habitat. In this perspective, any clade-specific gene family has potentially been involved in the stable adaptation of the clade to a new ecological niche. Therefore, new genes enabling the exploitation of a new ecological niche in a clade ancestor can be fixed and become core genes of the descendant taxon. Altogether, this suggests that pan- and core genomes are not isolated evolutionary compartments because core genes can be lost and variable genes can become essential. In fact, analysis of cellular networks shows a continuum of connectivity properties correlating with gene ubiquity, suggesting that new genes are progressively integrated into the network and the genome. It has been shown that after its acquisition, a gene slowly acquires more and more cis-regulatory elements, improves its co-expression with functional partners and progressively builds interactions with other gene products (Lercher and Pal, 2008). Notably, as newly acquired genes involved in ecological adaptation form interactions with core genes, they may become essential. *In fine*, the roles assumed by clade-specific genes involved in niche adaptation could become as essential as the replication or translation machinery in defining a bacterium's biology.

Hence, describing the ecological niche of prokaryotes, and particularly identifying to what extent genes acquired by HGT can modify its frontiers seems crucial to understand the dynamics of evolution of gene repertoires. While it is likely that species pangenomes contain a majority of genes transiently integrated in genomes with no particular adaptive fate, a fraction may support the ecological innovation of ecovar lineages, potentially leading to speciation (Cohan, 2002). If so, Cohan (2002) predicts that genes responsible for niche-specific adaptation of the descendant species will be kept in genomes of species members by drastic periodic selection, thus shifting from the unstable pangenome to the new species core genome. Actually, in studies characterizing species pan- and core genomes, many core genes have still no functional annotation, and are thus unlikely to participate to the rather well described bacterial central metabolism. Instead, they may participate in the adaptation of the species in its natural environment. Efforts must be made to phenotypically characterize those core genes and better understand their role in a species' biology. Novel approaches may consist of 'reverse ecology' in three steps: (1) genomic gene content comparison and gene phylogeny within a given taxon to identify clade-specific genes; (2) characterization of their function by bioinformatic and molecular biological methods; (3) integration of all functional information on niche-specific genes into a model of the taxon's ecological niche.

References

Achtman, M., and Wagner, M. (2008). Microbial diversity and the genetic nature of microbial species. Nat. Rev. Microbio. *6*, 431–440.

Angly, F.E., Willner, D., Prieto-Davó, A., Edwards, R.A., Schmieder, R., Vega-Thurber, R., Antonopoulos, D.A., Barott, K., Cottrell, M.T., Desnues, C., *et al.* (2009). The GAAS metagenomic tool and its estimations of viral and microbial average genome size in four major biomes. PLoS Comput. Biol. *5*, e1000593.

Charlebois, R.L., and Doolittle, W.F. (2004). Computing prokaryotic gene ubiquity: rescuing the core from extinction. Genome Res. *14*, 2469–2477.

Cohan, F.M. (2002). What are bacterial species? Annu. Rev. Microbiol. *56*, 457–487.

Cortez, D., Forterre, P., and Gribaldo, S. (2009). A hidden reservoir of integrative elements is the major source of recently acquired foreign genes and ORFans in archaeal and bacterial genomes. Genome Biol. *10*, R65.

Couturier, E., and Rocha, E.P.C. (2006). Replication-associated gene dosage effects shape the genomes of fast-growing bacteria but only for transcription and translation genes. Mol. Microbiol. *59*, 1506–1518.

Daubin, V., and Ochman, H. (2004). Bacterial genomes as new gene homes: the genealogy of ORFans in *E. coli*. Genome Res. *14*, 1036–1042.

Daubin, V., and Perrière, G. (2003). G+C3 Structuring along the genome: a common feature in prokaryotes. Mol. Biol. Evol. *20*, 471–483.

Davids, W., and Zhang, Z. (2008). The impact of horizontal gene transfer in shaping operons and protein interaction networks – direct evidence of preferential attachment. BMC Evol. Biol. *8*, 23.

Edwards, R.A., and Rohwer, F. (2005). Viral metagenomics. Nat. Rev. Microbiol. *3*, 504–510.

Esnault, E., Valens, M., Espéli, O., and Boccard, F. (2007). Chromosome structuring limits genome plasticity in *Escherichia coli*. PLoS Genet. *3*, e226.

Gerdes, S.Y., Scholle, M.D., Campbell, J.W., Balazsi, G., Ravasz, E., Daugherty, M.D., Somera, A.L., Kyrpides, N.C., Anderson, I., Gelfand, M.S., *et al.* (2003). Experimental determination and system level analysis of essential genes in *Escherichia coli* MG1655. J. Bacteriol. *185*, 5673–5684.

González, V., Bustos, P., Ramírez-Romero, M.A., Medrano-Soto, A., Salgado, H., Hernández-González, I., Hernández-Celis, J.C., Quintero, V., Moreno-Hagelsieb, G., Girard, L., *et al.* (2003). The mosaic structure of the symbiotic plasmid of Rhizobium etli CFN42 and its relation to other symbiotic genome compartments. Genome Biol. *4*, R36-R36.

Hao, W., and Golding, G.B. (2006). The fate of laterally transferred genes: life in the fast lane to adaptation or death. Genome Res. *16*, 636–643.

Hershberg, R., and Petrov, D.A. (2010). Evidence that mutation is universally biased towards AT in bacteria. PLoS Genet. *6*, e1001115.

Hildebrand, F., Meyer, A., and Eyre-Walker, A. (2010). Evidence of selection upon genomic GC-content in bacteria. PLoS Genet. *6*, pii: e1001107.

Kobayashi, K., Ehrlich, S.D., Albertini, A., Amati, G., Andersen, K.K., Arnaud, M., Asai, K., Ashikaga, S., Aymerich, S., Bessieres, P., *et al.* (2003). Essential *Bacillus subtilis* genes. Proc. Natl. Acad. Sci. U.S.A *100*, 4678–4683.

Koonin, E.V., and Wolf, Y.I. (2008). Genomics of bacteria and archaea: the emerging dynamic view of the prokaryotic world. Nucl. Acids Res. *36*, 6688–6719.

Kunin, V., Goldovsky, L., Darzentas, N., and Ouzounis, C.A. (2005). The net of life: reconstructing the microbial phylogenetic network. Genome Res. *15*, 954–959.

Kuo, C.-H., and Ochman, H. (2009). Deletional bias across the three domains of life. Genome Biol. Evol. *2009*, 145–152.

Lapierre, P., and Gogarten, J.P. (2009). Estimating the size of the bacterial pan-genome. Trends Genet. *25*, 107–110.

Lathe, W.C., Snel, B., and Bork, P. (2000). Gene context conservation of a higher order than operons. Trends Biochem. Sci. *25*, 474–479.

Lawrence, J.G., and Ochman, H. (1997). Amelioration of bacterial genomes: rates of change and exchange. J. Mol. Evol. *44*, 383–397.

Lefébure, T., Pavinski Bitar, P.D., Suzuki, H., and Stanhope, M.J. (2010). Evolutionary dynamics of complete *Campylobacter* pan-genomes and the bacterial species concept. Genome Biol. Evol.*2*, 646–655.

Lerat, E., Daubin, V., Ochman, H., and Moran, N.A. (2005). Evolutionary origins of genomic repertoires in bacteria. PLoS Biol. *3*, e130.

Lercher, M.J., and Pal, C. (2008). Integration of horizontally transferred genes into regulatory interaction networks takes many million years. Mol. Biol. Evol. *25*, 559–567.

Lobry, J.R., and Sueoka, N. (2002). Asymmetric directional mutation pressures in bacteria. Genome Biol. *3*, RESEARCH0058.

Marri, P.R., Hao, W., and Golding, G.B. (2007). The role of laterally transferred genes in adaptive evolution. BMC Evol. Biol. *7*, S8-S8.

Médigue, C., Rouxel, T., Vigier, P., Hénaut, A., and Danchin, A. (1991). Evidence for horizontal gene transfer in *Escherichia coli* speciation. J. Mol. Biol. *222*, 851–856.

Mira, A., Ochman, H., and Moran, N.A. (2001). Deletional bias and the evolution of bacterial genomes. Trends Genet.*17*, 589–596.

Mirkin, B., Fenner, T., Galperin, M., and Koonin, E. (2003). Algorithms for computing parsimonious evolutionary scenarios for genome evolution, the last universal common ancestor and dominance of horizontal gene transfer in the evolution of prokaryotes. BMC Evol. Biol. 3, 2.

Mizoguchi, H., Mori, H., and Fujio, T. (2007). *Escherichia coli* minimum genome factory. Biotechnol. Appl. Biochem. *46*, 157.

Mizoguchi, H., Sawano, Y., Kato, J., and Mori, H. (2008). Superpositioning of deletions promotes growth of *Escherichia coli* with a reduced genome. DNA Research *15*, 277–284.

Moran, N.A., and Plague, G.R. (2004). Genomic changes following host restriction in bacteria. Curr. Opin. Genet. Dev. *14*, 627–633.

Muller, J., Szklarczyk, D., Julien, P., Letunic, I., Roth, A., Kuhn, M., Powell, S., von Mering, C., Doerks, T., Jensen, L.J., *et al.* (2010). eggNOG v2.0: extending the evolutionary genealogy of genes with enhanced non-supervised orthologous groups, species and functional annotations. Nucleic Acids Res. *38*, D190–195.

Myllykallio, H., Lopez, P., Lopez-Garcia, P., Heilig, R., Saurin, W., Zivanovic, Y., Philippe, H., and Forterre, P. (2000). Bacterial mode of replication with eukaryotic-like machinery in a hyperthermophilic archaeon. Science *288*, 2212–2215.

Normand, P., Lapierre, P., Tisa, L.S., Gogarten, J.P., Alloisio, N., Bagnarol, E., Bassi, C.A., Berry, A.M., Bickhart, D.M., Choisne, N., *et al.* (2007). Genome characteristics of facultatively symbiotic Frankia sp. strains reflect host range and host plant biogeography. Genome Res. *17*, 7–15.

Ochman, H., and Davalos, L.M. (2006). The nature and dynamics of bacterial genomes. Science *311*, 1730–1733.

Ochman, H., and Moran, N.A. (2001). Genes lost and genes found: evolution of bacterial pathogenesis and symbiosis. Science *292*, 1096–1099.

Pal, C., Papp, B., and Lercher, M.J. (2005). Adaptive evolution of bacterial metabolic networks by horizontal gene transfer. Nat. Genet. *37*, 1372–1375.

van Passel, M.W.J., Marri, P.R., and Ochman, H. (2008). The emergence and fate of horizontally acquired genes in *Escherichia coli*. PLoS Comput. Biol. *4*.

Penel, S., Arigon, A.-M., Dufayard, J.-F., Sertier, A.-S., Daubin, V., Duret, L., Gouy, M., and Perriere, G. (2009). Databases of homologous gene families for comparative genomics. BMC Bioinformatics *10*, S3.

Porwollik, S., Wong, R.M.-Y., and McClelland, M. (2002). Evolutionary genomics of *Salmonella*: Gene acquisitions revealed by microarray analysis. Proc. Natl. Acad. Sci. U.S.A. *99*, 8956–8961.

Pósfai, G., Plunkett, G., 3rd, Fehér, T., Frisch, D., Keil, G.M., Umenhoffer, K., Kolisnychenko, V., Stahl, B., Sharma, S.S., de Arruda, M., *et al.* (2006). Emergent properties of reduced-genome *Escherichia coli*. Science *312*, 1044–1046.

Rahman, M.H., Biswas, K., Hossain, M.A., Sack, R.B., Mekalanos, J.J., and Faruque, S.M. (2008). Distribution of genes for virulence and ecological fitness among diverse *Vibrio cholerae* population in a cholera endemic area: tracking the evolution of pathogenic strains. DNA Cell Biol. 27, 347–355.

Rocha, E.P.C., and Danchin, A. (2003). Gene essentiality determines chromosome organisation in bacteria. Nucleic Acids Res. *31*, 6570–6577.

Rocha, E.P.C. (2006). Inference and analysis of the relative stability of bacterial chromosomes. Mol. Biol. Evol.*23*, 513–522.

Rocha, E.P.C. (2008). The organization of the bacterial genome. Genetics *42*, 211–233.

Snel, B., Bork, P., and Huynen, M.A. (2002). Genomes in flux: the evolution of archaeal and proteobacterial gene content. Genome Res. *12*, 17–25.

Tettelin, H., Masignani, V., Cieslewicz, M.J., Donati, C., Medini, D., Ward, N.L., Angiuoli, S.V., Crabtree, J., Jones, A.L., Durkin, A.S., *et al.* (2005). Genome analysis of multiple pathogenic isolates of *Streptococcus agalactiae*: Implications for the microbial 'pan-genome.' Proc. Natl. Acad. Sci. U.S.A. *102*, 13950–13955.

Touchon, M., Hoede, C., Tenaillon, O., Barbe, V., Baeriswyl, S., Bidet, P., Bingen, E., Bonacorsi, S., Bouchier, C., Bouvet, O., *et al.* (2009). Organised genome dynamics in the *Escherichia coli* species results in highly diverse adaptive paths. PLoS Genet. *5*, e1000344.

Treangen, T.J., and Rocha, E.P.C. (2011). Horizontal transfer, not duplication, drives the expansion of protein families in prokaryotes. PLoS Genet. *7*, e1001284.

Ussery, D.W., Kiil, K., Lagesen, K., Sicheritz-Pontén, T., Bohlin, J., and Wassenaar, T.M. (2009). The genus burkholderia: analysis of 56 genomic sequences. Genome Dyn *6*, 140–157.

Vesth, T., Wassenaar, T.M., Hallin, P.F., Snipen, L., Lagesen, K., and Ussery, D.W. (2010). On the origins of a *Vibrio* species. Microb. Ecol *59*, 1–13.

Williamson, S.J., Rusch, D.B., Yooseph, S., Halpern, A.L., Heidelberg, K.B., Glass, J.I., Andrews-Pfannkoch, C., Fadrosh, D., Miller, C.S., Sutton, G., *et al.* (2008). The Sorcerer II Global Ocean Sampling Expedition: metagenomic characterization of viruses within aquatic microbial samples. PLoS One *3*, e1456.

Woese, C.R., and Fox, G.E. (1977). Phylogenetic structure of the prokaryotic domain: the primary kingdoms. Proc. Natl. Acad. Sci. *74*, 5088–5090.

Yin, Y., and Fischer, D. (2008). Identification and investigation of ORFans in the viral world. BMC Genomics *9*, 24.

Yomtovian, I., Teerakulkittipong, N., Lee, B., Moult, J., and Unger, R. (2010). Composition bias and the origin of ORFan genes. Bioinformatics *26*, 996–999.

Horizontal Transfer of Host-adaptability Systems in Bacteria

3

Eva C. Berglund and A. Carolin Frank

Abstract

Understanding the origin, evolution, maintenance, and breakdown of host-adaptation is central to diverse areas such as human health and agriculture. Horizontal gene transfer (HGT) is a recurrent theme in the evolution of host-adaptability systems. Why are these genes often successfully transferred and what is the consequence for the recipient genome? Drawing examples from the mammal-adapted bacterium *Bartonella*, and from plant-adapted rhizobia, we discuss the role of horizontal gene transfer in the evolution and ecology of host-adaptation.

Introduction

Association with bacteria is a universal condition of large multicellular eukaryotes. Most bacteria, if given access to the forbidden niche inside eukaryotes, would probably be detrimental to the host, and eukaryotes have evolved effective strategies to keep the majority of microbes out. Here, we define *host-adapted* bacteria as those equipped with mechanisms to invade the eukaryotic niche and counter this defence, allowing them to persistently colonize eukaryote host tissues or cells. We further define *host-adaptability systems* as the genetic determinants underlying such abilities.

Many genes involved in host-interaction were first identified in disease-causing organisms, and are often described as 'pathogenicity' or 'virulence' determinants. However, it is important to remember that the distinction between pathogenic, mutualistic and commensal bacteria is relevant only from the host's perspective. Genes that mediate host-interaction have evolved to facilitate persistent colonization, not to cause disease, and they are typically shared among host-associated bacteria with different lifestyles. In this review, we address the role of host-adaptability gene transfer, without making distinctions among the outcome from the perspective of the host. Diverse bacteria–eukaryote interactions are found in the Rhizobiales order of Alphaproteobacteria. We focus here on the blood-borne bacterium *Bartonella* and its mammal hosts, and also on the interaction between rhizobia and their plant host. We begin with a brief background on host–microbe relationships and their genomic basis, followed by a section on *Bartonella*. We then explore the role of HGT in catalysing and refining host–microbe interactions, and discuss the properties of host-interaction traits that make them amenable to horizontal transfer. Finally, we consider in detail the combination of a phage run-off replication and a gene transfer agent as a possible adaptation to maintain variability of host-interaction genes in *Bartonella*.

Host–microbe relationships

Multicellular eukaryotes engage in interaction with a broad range of bacteria with different transmission dynamics, niche within the host, and ability to survive outside the host (Toft and Andersson, 2010). Vertically transmitted bacteria are inherited from parent to offspring, whereas horizontally transmitted bacteria are acquired from the environment or from other host individuals. Facultatively host-associated bacteria have a free-living life stage outside of their host, whereas obligate host-associated bacteria do not [note that this is with respect to the bacterium – one can also talk about facultative or obligate mutualists, but from the host's perspective (Oliver *et al.*, 2010)]. Examples of facultative host-associated bacteria are the rhizobia (nodule-forming nitrogen-fixing plant symbionts), which can live and replicate free-living in soil as well as in association with plants. Other plant-associated bacteria, such as some nitrogen-fixing endophytes of grasses (e.g. *Gluconobacter diazotrophicus*), cannot be isolated from soil and are believed to replicate only in host plant tissue, i.e. they are obligate host-associated (Kaneko *et al.*, 2010). Finally, some host-associated bacteria can colonize the inside of eukaryote cells, either obligatorily (when the bacterium can only persist and replicate within hosts cells) or facultatively (when the bacterium can persist and replicate inside or outside of host cells). Host-association or intracellularity is often classified as facultative if bacteria can be cultured *in vitro*, even for bacteria that do not have a free-living or extracellular stage in their natural life cycle. The aphid endosymbiont *Buchnera* is an obligate intracellular, vertically transmitted bacterium that cannot be cultured *in vitro*, with no part of its life cycle outside of the host. In contrast, the vector-borne, mammal-associated bacterium *Bartonella*, from which we draw many of our examples in this chapter, replicates intracellularly in red blood cells. *Bartonella* is often described as a facultative intracellular bacterium – although it does not naturally grow outside its mammal host or insect or arthropod vectors, it can be cultivated on blood agar (Regnery *et al.*, 1992).

Genomic basis of host-adaptability

The adaptation to a host-associated lifestyle is reflected in the bacterial genome. With the exception of obligate, vertically transmitted bacteria that do not depend on host exit and re-entry, a typical feature of host-adaptation is the presence of genes involved in host surface- or cell attachment and invasion, and host defence evasion or subversion. In addition, many host-adapted bacteria have evolved mechanisms to modulate host biology to extend their niche, e.g. through the proliferation of infected host cells (Batut *et al.*, 2004) or through mechanisms that enhance their survival while benefiting the host, such as symbiotic nitrogen fixation (West *et al.*, 2002). Here, we focus specifically on secretion systems, which are common facilitators of bacteria–eukaryote associations (Meyer and Rosenshine, 2008). Secretion systems are multiprotein machineries that translocate molecules called effectors across the cell membrane(s), from the bacterial cell cytoplasm into the extracellular milieu, or directly into the eukaryotic host cell (Tseng *et al.*, 2009). There are currently seven known distinct non-general secretion systems (designated type I through type VII), implicated in diverse processes such as invasion, attachment, defence evasion, host cell modulation, niche extension and host range modulation. The majority of secretion systems deliver a molecule (protein or DNA) that modulates the interaction with the host, but some, such as *Bartonella* Trw, are directly involved in host cell attachment (Vayssier-Taussat *et al.*, 2010). Secretion systems are not, as previously assumed, unique to disease-causing organisms, but typically shared among host-associated bacteria with different lifestyles (Coombes, 2009; Deakin and Broughton, 2009; Jani and Cotter, 2010). For example, secretion systems and their effectors are important host-adaptability systems in both mutualistic, parasitic and non-symbiotic rhizobia (Fauvart and Michiels, 2008).

The genus *Bartonella*: host adaptive radiation

Bartonella is a genus of facultative intracellular bacteria that colonize erythrocytes and endothelial cells of mammals. *Bartonella* spp. do not replicate outside their hosts, or the blood-sucking insect and arthropod vectors, that they use for transmission. The genus *Bartonella* belongs to the Rhizobiales order of Alphaproteobacteria. Many Rhizobiales species are plant-adapted, such as the plant-pathogen *Agrobacterium tumefaciens* and the rhizobia *Mesorhizobium loti* and *Sinorhizobium meliloti*. *Bartonella*'s closest relatives outside of the genus are the animal-associated *Brucella* spp., and *Ochrobactrum* spp. which are found in a wide range of environments, including plants (as rhizobia) (Ngom *et al.*, 2004), and opportunistically in animal tissues, including blood (Kampfer *et al.*, 2007). There are more than 20 described species of *Bartonella*, and they infect a wide range of wild and domestic mammals, including humans, felines, rodents, ruminants and canines. The genus *Bartonella* has undergone adaptive radiation in concert with host species, and most *Bartonella* species are restricted to one or a group of closely related species of mammal hosts called the natural host. Switching of the natural host appears to be rare. Most *Bartonella* species have evolved with their host over long periods of time, and do not cause symptoms. Incidental cross-species infection on the other hand is common and can lead to disease. For example, human infection with the feline-adapted *Bartonella henselae* causes cat-scratch disease. *Bartonella quintana*, which is the agent of trench fever and affected millions of soldiers during World War I (Bass *et al.*, 1997), is related to the feline-associated *B. henselae*, but is strictly human-associated. *B. quintana* is believed to have undergone an evolutionary natural host switch from felines to humans (Alsmark *et al.*, 2004). *Bartonella* colonizes and replicates within erythrocytes of their natural hosts but not in incidentally infected species. Intra-erythrocytic infections are host-restricted, and host-specificity is determined by the capacity of bacteria to adhere to erythrocytes, mediated by the type IV secretion system Trw (Vayssier-Taussat *et al.*, 2010). *Bartonella* spp. also colonize a primary niche (probably endothelial cells), which seeds bacteria into the blood stream.

The genomes of ten *Bartonella* species have been sequenced to date (Alsmark *et al.*, 2004; Saenz *et al.*, 2007; Berglund *et al.*, 2009; Engel *et al.*, 2011). They all consist of a single circular chromosome between 1.4 and 2.6 Mb in size. The *Bartonella* genomes are in general well conserved, and appear to be reduced versions of chromosome I of the closely related *Brucella* species. However, a 2–500 kb segment of the *Bartonella* genomes displays higher similarity to chromosome II of *Brucella*, has a lower coding content and a lower degree of synteny relative to the rest of the genome (Alsmark *et al.*, 2004). Because of its atypical characteristics, this segment is called the chromosome II-like region or the chromosomal plasticity region. This region harbours many of the identified host-adaptability genes, such as the type IV secretion systems VirB and Trw, the flagellum, and several families of type V secretion systems (Berglund *et al.*, 2009).

Horizontal transfer and the origin of bacteria–host interactions

Phylogenetic distribution of host-adaptation

Several lines of evidence suggest that the successful horizontal transfer of host-adaptability genes among genomes is common, and that this transfer contributes to the evolution of host-adaptability. First, although bacteria from some phyla, such as the Proteobacteria, have been more successful in adapting to the eukaryotic niche than others (Toft and Andersson, 2010), host-adapted bacteria are also found in distant phyla across the bacterial tree such as Tenericutes (e.g. *Mycoplasma* spp.), Firmicutes (e.g. *Listeria* spp.), Bacteroidetes (e.g. *Blattabacterium* spp.), Actinobacteria (e.g. *Frankia* spp.) and Cyanobacteria (e.g. *Nostoc* spp.). Therefore, host-adaption in these lineages must have evolved independently. In some instances, origin of host-adaptation may have been facilitated by horizontal transfer of host-interaction traits. For example, rhizobial plant symbionts occur both in the Alpha- and Betaproteobacteria (Moulin *et al.*, 2001; Masson-Boivin *et al.*, 2009) and actinorhizal symbionts occur in the *Frankia* genus of Actinobacteria (Benson, 1988; Clawson *et al.*, 2004), yet, they all use the same highly conserved enzymes to fix atmospheric nitrogen (Raymond *et al.*, 2004). Second, the phylogenetic distribution of host-adapted bacteria is often discontinuous within lower taxonomic ranks such as genera, suggesting that host-adaptation correlates with the presence of genes that are absent from non-host-adapted relatives. For example, many members of the genus *Methylobacterium* live in soil, air and water environments (Lidstrom, 2006), whereas others are engaged in endophytic or root nodule forming symbiosis with plants (Pirttilä *et al.*, 2000; Madhaiyan *et al.*, 2009). Likewise, symbiotic nitrogen fixation is sporadic within the Rhizobiales, and is thought to have originated from independent events of horizontal transfer of key symbiotic functions (*nod, nif* and *fix* genes, responsible for invasion/nodulation and nitrogen fixation respectively) from symbiotic to non-symbiotic genomes (Sullivan *et al.*, 1995; Sullivan and Ronson, 1998; Martinez-Romero, 2009; Masson-Boivin *et al.*, 2009). Absence/presence patterns of these genes correlate with different lifestyles even among strains within a single species (*Rhizobium leguminosarum*) (Denison and Kiers, 2004).

Transferability of host-adaptability systems

Not all genes are equally likely to be transferred horizontally from a donor genome or retained in a recipient genome. Two common qualities of host-adaptability genes make them amenable to transfer: they are rarely part of large protein interaction networks, and they are often associated with mobile genetic elements. According to the complexity hypothesis, genes involved in informational processes and more highly connected in protein–protein interactions networks, are less likely to be successfully transferred than metabolic (operational) genes, which have a lower level of connectivity (Jain *et al.*, 1999; Cohen *et al.*, 2011). Many host-adaptability systems fall in the operational category, i.e. they do not play a role in information processes and they are not part of large, complex systems. For example, while the proteins in a secretion system interact among themselves (Shamaei-Tousi *et al.*, 2004), they typically do not form interactions with other proteins in the cell that would otherwise create a barrier to their successful transfer. Moreover, secretion systems are promiscuous with regards to the effectors they secrete (Anderson *et al.*, 1999), even to the point where different secretion systems can transport the same effectors (Hubber *et al.*, 2004). Finally, unlike the secretion systems themselves, secreted effectors are not conserved among genomes (Stavrinides *et al.*, 2008). Thus, even if a secretion system is transferred without its effectors, new secretion system–effector protein relationships can probably evolve relatively easily.

In many species, host-adaptability factors are part of the accessory genome, and they tend to be associated with mobile genetic elements. This is consistent with the observation that mobile elements frequently evolve to confer benefits on their bacterial hosts (Rankin *et al.*, 2011). For example, rhizobial *nod, nif* and *fix* genes tend to be located in genomic islands called 'symbiosis islands', or on large plasmids called pSym (MacLean *et al.*, 2007). The symbiosis island in *M. loti* is an integrative and conjugative element that excises from the genome as a circular intermediate that can be transferred by conjugation from symbiotic to non-symbiotic strains (Sullivan *et al.*, 1995; Sullivan and Ronson, 1998; Ramsay *et al.*, 2006). In most rhizobia however, *nod, nif* and *fix* genes are encoded on pSym plasmids (MacLean *et al.*, 2007), which are commonly transferred among bacteria through conjugation (Ding and Hynes, 2009).

Horizontal transfer of secretion systems

In rhizobia as well as other bacteria, secretion systems tend to be part of the accessory genome (Deakin and Broughton, 2009), and appear to move easily across species boundaries (Frank *et al.*, 2005; Ma *et al.*, 2006; McCann and Guttman, 2008). Interestingly, several secretion systems (type III, IV and VI) appear to have been co-opted from functions that likely existed before the origin of multicellular eukaryotes (Frank *et al.*, 2005; McCann and Guttman, 2008; Leiman *et al.*, 2009). Especially the co-option of bacterial conjugation for effector translocation has happened many times independently in different bacterial species, often in association with horizontal transfer of the system into a new genetic background (Frank *et al.*, 2005). Since conjugation systems encode the machinery for their own transfer, it is not surprising that their horizontal transfer is common.

In *Bartonella*, two type IV secretion systems have been imported and co-opted for use in different stages of the infection process, and these acquisitions were key to the evolution of the genus. *Bartonella* VirB is most closely related to a group of VirB systems found in plant-associated Rhizobiales species [e.g. the *Agrobacterium tumefaciens* VirB that is encoded on the Ti-plasmids, and delivers DNA to the plant host cell (Kado, 1994)]. It translocates a mix of proteins into endothelial host cells (the proposed primary niche), where they inhibit apoptosis (Schmid *et al.*, 2006) and trigger uptake of large bacterial aggregates (Rhomberg *et al.*, 2009). Trw, the other *Bartonella* type IV secretion system, does not deliver effectors but has been co-opted for a different type of host-adaptability function. The pilus components of Trw mediate binding to structures on the surface of red blood cells, and are required for erythrocyte invasion (Vayssier-Taussat *et al.*, 2010). The operon is only present in the most species-rich of four *Bartonella* lineages (lineage 1, where *B. henseale, B. quintana* and *Bartonella grahamii,* which lives in rodents and is further discussed below, belong) (Engel *et al.*, 2011). *Bartonella trw* is highly similar to the *Escherichia coli* plasmid R388 *trw* conjugation system operon (Seubert *et al.*, 2003; Frank *et al.*, 2005), suggesting that it was transferred to *Bartonella* relatively recently. Alternatively, *trw* may have been acquired in the ancestor of all *Bartonella* species and subsequently lost in lineages 2, 3 and 4. The presence of *trw* in lineage 1 correlates with the absence of a flagellum-encoding operon. Flagella are required for the invasion of erythrocytes by *B. bacilliformis* and probably other *Bartonella* species encoding them (Vayssier-Taussat *et al.*, 2010).

Recombination and within-population variability of host-adaptability systems

Rapid sequence evolution is a common feature of host-adaptability genes. Bacteria that depend on host invasion need to keep up with a changing and variable environment created by host defence mechanisms and host surface structures. Recombination-mediated dynamic variability of genes can serve as a means to maintain robustness against such challenges. However, an overall elevated recombination rate would potentially interfere with informational genes and genes that are highly connected in protein–protein interaction networks (Jain *et al.*, 1999). It is possible that bacteria have evolved strategies that allow them to generate variation locally (e.g. in specific genes, operons, replicons or genomic regions), while keeping recombination rates in other parts of the genome down. As discussed below, higher recombination rates are documented within host-adaptability operons, in additional replicons, and in the case of *Bartonella,* in a specific genomic region (the chromosomal plasticity region) in which the combined action of run-off replication and a gene transfer agent may be an adaptive strategy to selectively create genetic variation in host-adaptability systems.

Rapid recombination of single genes

Locally high rates of recombination drive rapid evolution of *Bartonella* Trw, the host-adaptability system used in erythrocyte adherence (Vayssier-Taussat *et al.*, 2010). Erythrocyte surface structures such as glycophorines are among the fastest evolving genes in the human genome, probably as a result of intense selective pressure to evade erythrocyte parasites such as *Plasmodium* and other haemosporidians (Wang *et al.*, 2003). To efficiently attach to and invade erythrocytes, *Bartonella* needs to match this rapid evolution. In all *Bartonella* species that carry this operon (lineage 1), two segments, both containing genes encoding surface exposed proteins, are duplicated. The gene encoding the major component of the pilus, *trw*L, is present in seven or eight copies, and a segment that includes *trw*J, which mediates host cell attachment, is present in variable copy number in the different species. These duplicated *trw* genes evolve extremely rapidly with frequent indels through extensive intra- and intergenome recombination (Nystedt *et al.*, 2008). The precise mechanism behind this high local recombination rate is unknown. In some host-adapted species, such as *Anaplasma marginale,* one of several duplicated copies is expressed, whereas the others are silent and serve as sequence reservoirs that feed divergence into the expressed site (Brayton *et al.*, 2002; Foley *et al.*, 2009). For *Bartonella trw,* however, there is no evidence supporting the existence of silent copies. Instead, horizontal transfer from distant strains might feed divergence into *trw*. For example, there is evidence of acquisition of a large portion of the *trwL* by a recent horizontal gene transfer into a *B. henselae* lineage from a rather distantly related strain.

Multiple replicons and recombination of host-adaptability genes

Around 1 in 10 bacteria have more than one large replicon (Egan *et al.*, 2005; Harrison *et al.*, 2010). Interestingly, most bacteria with multipartite genomes are host-adapted (Egan *et al.*, 2005), suggesting that divided genomes provide an advantage to bacteria that interact with eukaryotes. For example, many rhizobia have multipartite genomes (MacLean *et al.*, 2007). Secondary chromosomes contain large amounts of horizontally transferred genes, tend to carry adaptability functions, and undergo frequent recombination (Guo *et al.*, 2007; Ettema and Andersson, 2009; Bavishi *et al.*, 2010). However, it is unclear what selective advantages and mechanisms maintain divided genomes over evolutionary time (Egan *et al.*, 2005; Teyssier *et al.*, 2005). One potential explanation is that a multipartite genome increases a bacterium's capacity to adapt to a variable or challenging environment (Gonzalez *et al.*, 2006).

Secondary chromosomes are thought to have evolved from plasmids, and are generally believed to encode accessory functions used only in specific niches or conditions (Holden *et al.*, 2004; Ettema and Andersson, 2009). While main chromosomes typically have more conserved and essential housekeeping genes, higher gene synteny, and higher degree of gene content conservation (Heidelberg *et al.*, 2000; Holden *et al.*, 2004), secondary chromosomes experience more rapid evolution at both the gene content and sequence divergence level. Secondary chromosomes are highly variable even among strains from the same species (Giuntini *et al.*, 2005; Teyssier *et al.*, 2005). Orthologues on secondary chromosomes evolve more rapidly, possibly as a consequence of reduced expression and purifying selection (Cooper *et al.*, 2010). To maintain synchronous replication, a smaller chromosome may be replicated later, resulting in a reduced copy number and lower expression level of genes on that chromosome [similar to how the dosage and expression decreases with increased distance to the origin of replication (Sharp *et al.*, 1989)]. The resulting chromosome-specific expression levels may allow selection to alter the expression of individual genes by relocating them to a different chromosome (Cooper *et al.*, 2010). This is compatible with the idea that multipartite genomes increase the adaptive capacity of a genome through reassortment of essential, non-essential and redundant genes (Gonzalez *et al.*, 2006), and could explain the long-term segregation of genome content observed in certain bacterial genera. Cooper *et al.* (2010) suggest that secondary chromosomes are 'evolutionary test beds' subject to faster evolution. A similar argument can be made for the observed rapid gene turnover on secondary chromosomes. Physical separation of informational or highly connected genes from operational/adaptability genes may be beneficial if secondary chromosomes are more prone to recombination. Indeed, secondary chromosomes are subject to common conjugative transfer, as suggested by both comparative genomics and experimentally documented transfers (Ding and Hynes, 2009). Additionally, differential expression could contribute to faster gene content evolution on secondary chromosomes if incoming genes may be relocated to the second chromosome to keep their expression level down. Regardless of the mechanisms underlying the stability of divided genomes, secondary replicons may facilitate HGT of host-adaptability traits, in turn refining host-adaptation, and generating variability within populations of host-adapted bacteria.

Gene transfer agents and run-off replication in *Bartonella*

Bartonella is a potential ancient derivative of plant-associated *Rhizobiales*. Although *Bartonella*'s closest known relatives, *Brucella* and *Ochrobactrum,* both have multipartite genomes, *Bartonella* has one single chromosome, however with remnants of an integrated second chromosome (or part of one). It is conceivable that like other host-adapted *Rhizobiales,* the ancestor of *Bartonella* carried its host-adaptability systems on a second chromosome. As *Bartonella* became host-restricted and these functions became indispensable, integration of the secondary replicon may have been selectively favoured to avoid the loss of important host-adaptability genes due to genetic drift. However, this event may also have limited the adaptive potential of *Bartonella.* The problem of reduced recombination following integration appears to have been solved by a peculiar set of events. A combination of a phage run-off replication and a gene transfer agent may allow a continuation of high local recombination rates in this genomic region, which carries many host-adaptability genes. As a consequence, this region experiences faster evolution relative to the rest of the genome, and is called the chromosomal plasticity region.

The discovery of the *Bartonella* gene transfer agent began with the observation of phage-like particles containing 14 kb linear, double-stranded DNA in several species (Umemori *et al.*, 1992; Anderson *et al.*, 1999; Barbian and Minnick, 2000; Maggi and Breitschwerdt, 2005). When the first *Bartonella* genomes were sequenced a decade later, it was discovered that many accessory and host-specific genes were clustered in a large region approximately corresponding to the fourth quarter of the chromosome (Alsmark *et al.*, 2004). This was apparent in *B. henselae*, but less so in the highly reduced human specialist and pathogen *B. quintana*. However, the sequences did not offer an explanation of the mysterious phage-particles, since there was no 14-kb phage present in the genomes. Despite several attempts to characterize the proteins making up the phage particles and their DNA content, these issues were not resolved until the sequencing of the *Bartonella grahamii* genome and the analysis of phages in this species through isolation and microarray hybridization of the 14 kb DNA band to the entire genome (Berglund *et al.*, 2009). Although the entire chromosome was represented in the 14 kb band, there was an increase in signal intensity over the region previously hypothesized to derive from an integrated secondary replicon (the chromosomal plasticity region). The microarray signal peaked within this region, and decreased gradually in both directions, suggesting bidirectional replication from an unknown origin and random termination, resulting in overamplification of the chromosomal plasticity region relative to the rest of the genome.

This pattern is similar to what has been observed in genomic DNA from several strains of *B. henselae* (Lindroos *et al.*, 2006), and to a smaller extent in other strains of *B. grahamii* (Berglund *et al.*, 2010a,b). In both species, the peak of the amplification is located in a short cluster of phage genes. Though at first glance this cluster appears to be a degraded prophage, it is actually well conserved across *Bartonella* genomes. Phage-induced replication of large amounts of bacterial DNA has also been observed in *Salmonella enterica* and *Escherichia coli* (Frye *et al.*, 2005; Asadulghani *et al.*, 2009), and has been termed *escape replication* or *run-off replication*. However, in *S. enterica* and *E. coli*, the origin of the run-off replication was located in a complete prophage. Moreover, in contrast to the *Bartonella* phage particles, which contain genomic DNA, analysis of one of the phages that induced escape replication in *S. enterica* showed that despite the abnormal replication it packages only its own DNA (Frye *et al.*, 2005).

A phage particle that packages random bacterial DNA instead of its own DNA is called a gene transfer agent (GTA) (Lang and Beatty, 2007). The process is similar to generalized transduction, but the GTA does not self-propagate. GTAs are typically encoded by a gene cluster of circa 15–20 kb, and are normally not associated with lysis or other negative consequences for the host. One of the most well studied GTAs is the RcGTA of *Rhodobacter capsulatus*, which is well conserved among the Alpha-proteobacteria (Leung *et al.*, 2010). Because of their wide distribution among Alphaproteobacteria, GTAs are generally assumed to be adaptive, conferring beneficial gene transfer to host populations (Lang and Beatty, 2007). However, the possibility that they are merely defective prophages cannot be ruled out. Regardless of the outcome of GTA action on host populations, it is clear that they contribute to massive amounts of gene transfer. A recent study of GTAs from *Roseovarius nubinhibens* and *Reugeria mobilis* revealed unprecedented frequencies of GTA-mediated horizontal gene transfer in marine environments (McDaniel *et al.*, 2010).

SDS-PAGE and mass spectrometry of isolated *B. grahamii* GTA particles allowed the identification of three proteins, all encoded in a gene cluster that comprises circa 15 kb and is well conserved in all sequenced *Bartonella* genomes (Berglund *et al.*, 2009). A few genes in this cluster encode recognized phage proteins, and the remaining are annotated as encoding hypothetical proteins, and lack similarity to previously described GTAs. However, the identification of three proteins encoded in this cluster in the GTAs, and the conservation in all species where phage particles have been observed, strongly suggests that this cluster corresponds to GTA genes. These genes are located within the amplified chromosomal plasticity region, however not close to the peak of the amplification.

While both GTAs and run-off replication have been observed previously in other species, the association of these two phenomena has to our knowledge only been demonstrated in *Bartonella*. An interesting question is whether this combination provides a selective advantage for the bacterial host, or merely is a peculiar effect resulting from two defective prophages ending up in the same genome. Since a GTA does not contain its own DNA and cannot propagate once it has entered a new host, it is likely to disappear from a bacterial genome unless it provides some benefit to the bacterial population. The high conservation of both the genes encoding the GTA and the genes located at the peak of the amplification suggests that the phenomenon is adaptive. What could then be the benefit of linking a GTA to run-off replication? While GTAs have been discussed in the context of adaptive evolution (Lang and Beatty, 2007) because of their ability to increase gene exchange and recombination, run-off replication has not. Recombination has indeed been demonstrated in several rodent-associated *Bartonella* species (Paziewska *et al.*, 2011), and the recombination rates of *B. grahamii* are very high in comparison to most intracellular bacteria (Berglund *et al.*, 2010b). Increased genetic diversity of genes that are surface-exposed and/or involved in host interaction processes can facilitate evasion of the host immune system or the binding to rapidly evolving host surface structures. There are however also drawbacks with recombination, since beneficial allele combinations can be disrupted. Thus, increasing the amount of recombination of genes involved in host interactions relative to housekeeping genes could potentially be a key to success. A closer inspection of the over-amplified region in *B. grahamii* shows that it contains many genes involved in host interactions, including the type IV secretion systems described above and several families of type V secretion systems. Run-off replication increases dosage of these genes, thus increasing the probability that they are packaged by the GTA, transferred and recombined. The combination of a GTA with run-off replication might thus be a means to increase the genetic diversity of host interaction genes, while reducing the general burden of recombination.

Conclusion

Association with, or exclusion of microbes is crucial to the health of most eukaryotes. Bacteria adapted to a lifestyle within a eukaryotic host have evolved independently in several phyla, and a genus or even a species of bacteria can contain both host-adapted and free-living strains. Horizontal gene transfer can contribute to the origin of new host–microbe associations and can refine existing host interactions. It is important to point out however, that instant host-adaptation through gene transfer is probably rare. It is more likely that the predecessors of host-adapted bacteria such as *Bartonella* or rhizobia had already explored the eukaryotic niche, perhaps as opportunists, and possessed some of the traits required to cope with the host environment, including mechanisms to counteract host defences. In other words, they were pre-adapted to become better adapted to hosts when they acquired additional host-adaptability factors. For example, when an ancestor of lineage 1 *Bartonella* acquired the *trw* system, it was already adapted to life in mammal blood (because lineage 1 *Bartonella* spp. shares this niche with other *Bartonella* lineages). However, the acquisition of *trw* probably allowed efficient binding to erythrocytes, matching the rapid evolution of erythrocyte surface structures. Also, additional changes may be required post acquisition of host-adaptability genes, in order to improve host-adaptation. For example, rhizobial symbiosis plasmid transfer alone is not enough to create a nodulating strain, but subsequent adaptive mutations are required (Marchetti *et al.*, 2010).

In order to persistently colonize and escape the immune system of a host, rapid sequence evolution of host adaptability genes is often required, and bacteria appear to have evolved several strategies for this purpose. Mechanisms for rapid evolution of host-adaptability traits often involve increased recombination rates, conjugation or transduction. The targeted regions range from single genes or gene families to entire secondary chromosomes. In the future it will be interesting to see if the potentially adaptive trait recently discovered in *Bartonella,* where the combination of two defective phages seems to increase recombination rates in a region of several hundred kb which is especially dense in host-adaptability genes, is as rare as it seems, or if similar phenomena will be discovered in other host-adapted bacteria.

References

Alsmark, C.M., Frank, A.C., Karlberg, E.O., Legault, B.A., Ardell, D.H., Canback, B., Eriksson, A.S., Naslund, A.K., Handley, S.A., Huvet, M., *et al.* (2004). The louse-borne human pathogen *Bartonella quintana* is a genomic derivative of the zoonotic agent *Bartonella henselae*. Proc. Natl. Acad. Sci. U.S.A. *101*, 9716–9721.

Anderson, D.M., Fouts, D.E., Collmer, A., and Schneewind, O. (1999). Reciprocal secretion of proteins by the bacterial type III machines of plant and animal pathogens suggests universal recognition of mRNA targeting signals. Proc. Natl. Acad. Sci. U.S.A. *96*, 12839–12843.

Asadulghani, M., Ogura, Y., Ooka, T., Itoh, T., Sawaguchi, A., Iguchi, A., Nakayama, K., and Hayashi, T. (2009). The defective prophage pool of *Escherichia coli* O157: prophage–prophage interactions potentiate horizontal transfer of virulence determinants. PLoS Pathog. *5*, e1000408.

Barbian, K.D., and Minnick, M.F. (2000). A bacteriophage-like particle from *Bartonella bacilliformis*. Microbiology *146*, 599–609.

Bass, J.W., Vincent, J.M., and Person, D.A. (1997). The expanding spectrum of *Bartonella* infections: I. Bartonellosis and trench fever. Pediatr. Infect. Dis. J. *16*, 2–10.

Batut, J., Andersson, S.G., and O'Callaghan, D. (2004). The evolution of chronic infection strategies in the alpha-proteobacteria. Nat. Rev. Microbiol. *2*, 933–945.

Bavishi, A., Abhishek, A., Lin, L., and Choudhary, M. (2010). Complex prokaryotic genome structure: rapid evolution of chromosome II. Genome *53*, 675–687.

Benson, D.R. (1988). The genus Frankia: actinomycete symbionts of plants. Microbiol. Sci. *5*, 9–12.

Berglund, E.C., Ehrenborg, C., Vinnere Pettersson, O., Granberg, F., Naslund, K., Holmberg, M., and Andersson, S.G. (2010a). Genome dynamics of *Bartonella grahamii* in micro-populations of woodland rodents. BMC Genomics *11*, 152.

Berglund, E.C., Ellegaard, K., Granberg, F., Xie, Z., Maruyama, S., Kosoy, M.Y., Birtles, R.J., and Andersson, S.G. (2010b). Rapid diversification by recombination in *Bartonella grahamii* from wild rodents in Asia contrasts with low levels of genomic divergence in Northern Europe and America. Mol. Ecol. *19*, 2241–2255.

Berglund, E.C., Frank, A.C., Calteau, A., Vinnere Pettersson, O., Granberg, F., Eriksson, A.S., Naslund, K., Holmberg, M., Lindroos, H., and Andersson, S.G. (2009). Run-off replication of host-adaptability genes is associated with gene transfer agents in the genome of mouse-infecting *Bartonella grahamii*. PLoS Genet. *5*, e1000546.

Brayton, K.A., Palmer, G.H., Lundgren, A., Yi, J., and Barbet, A.F. (2002). Antigenic variation of *Anaplasma marginale* msp2 occurs by combinatorial gene conversion. Mol. Microbiol. *43*, 1151–1159.

Clawson, M.L., Bourret, A., and Benson, D.R. (2004). Assessing the phylogeny of Frankia-actinorhizal plant nitrogen-fixing root nodule symbioses with Frankia 16S rRNA and glutamine synthetase gene sequences. Mol. Phylogenet. Evol. *31*, 131–138.

Cohen, O., Gophna, U., and Pupko, T. (2011). The complexity hypothesis revisited: connectivity rather than function constitutes a barrier to horizontal gene transfer. Mol. Biol. Evol. *28*, 1481–1489.

Coombes, B.K. (2009). Type III secretion systems in symbiotic adaptation of pathogenic and non-pathogenic bacteria. Trends Microbiol. *17*, 89–94.

Cooper, V.S., Vohr, S.H., Wrocklage, S.C., and Hatcher, P.J. (2010). Why genes evolve faster on secondary chromosomes in bacteria. PLoS Comput. Biol. *6*, e1000732.

Deakin, W.J., and Broughton, W.J. (2009). Symbiotic use of pathogenic strategies: rhizobial protein secretion systems. Nat. Rev. Microbiol. *7*, 312–320.

Denison, R.F., and Kiers, E.T. (2004). Lifestyle alternatives for rhizobia: mutualism, parasitism, and forgoing symbiosis. FEMS Microbiol. Lett. *237*, 187–193.

Ding, H., and Hynes, M.F. (2009). Plasmid transfer systems in the rhizobia. Can. J. Microbiol. *55*, 917–927.

Egan, E.S., Fogel, M.A., and Waldor, M.K. (2005). Divided genomes: negotiating the cell cycle in prokaryotes with multiple chromosomes. Mol. Microbiol. *56*, 1129–1138.

Engel, P., Salzburger, W., Liesch, M., Chang, C.C., Maruyama, S., Lanz, C., Calteau, A., Lajus, A., Medigue, C., Schuster, S.C., *et al.* (2011). Parallel evolution of a type IV secretion system in radiating lineages of the host-restricted bacterial pathogen *Bartonella*. PLoS Genet. *7*, e1001296.

Ettema, T.J., and Andersson, S.G. (2009). The alpha-proteobacteria: the Darwin finches of the bacterial world. Biol. Lett. *5*, 429–432.

Fauvart, M., and Michiels, J. (2008). Rhizobial secreted proteins as determinants of host specificity in the rhizobium–legume symbiosis. FEMS Microbiol. Lett. *285*, 1–9.

Foley, J.E., Nieto, N.C., Barbet, A., and Foley, P. (2009). Antigen diversity in the parasitic bacterium *Anaplasma phagocytophilum* arises from selectively-represented, spatially clustered functional pseudogenes. PLoS One *4*, e8265.

Frank, A.C., Alsmark, C.M., Thollesson, M., and Andersson, S.G. (2005). Functional divergence and horizontal transfer of type IV secretion systems. Mol. Biol. Evol. *22*, 1325–1336.

Frye, J.G., Porwollik, S., Blackmer, F., Cheng, P., and McClelland, M. (2005). Host gene expression changes and DNA amplification during temperate phage induction. J. Bacteriol. *187*, 1485–1492.

Giuntini, E., Mengoni, A., De Filippo, C., Cavalieri, D., Aubin-Horth, N., Landry, C.R., Becker, A., and Bazzicalupo, M. (2005). Large-scale genetic variation of the symbiosis-required megaplasmid pSymA revealed by comparative genomic analysis of *Sinorhizobium meliloti* natural strains. BMC Genomics *6*, 158.

Gonzalez, V., Santamaria, R.I., Bustos, P., Hernandez-Gonzalez, I., Medrano-Soto, A., Moreno-Hagelsieb, G., Janga, S.C., Ramirez, M.A., Jimenez-Jacinto, V., Collado-Vides, J., *et al.* (2006). The partitioned Rhizobium etli genome: genetic and metabolic redundancy in seven interacting replicons. Proc. Natl. Acad. Sci. U.S.A. *103*, 3834–3839.

Guo, X., Flores, M., Morales, L., Garcia, D., Bustos, P., Gonzalez, V., Palacios, R., and Davila, G. (2007). DNA diversification in two *Sinorhizobium* species. J. Bacteriol. *189*, 6474–6476.

Harrison, P.W., Lower, R.P., Kim, N.K., and Young, J.P. (2010). Introducing the bacterial 'chromid': not a chromosome, not a plasmid. Trends Microbiol. *18*, 141–148.

Heidelberg, J.F., Eisen, J.A., Nelson, W.C., Clayton, R.A., Gwinn, M.L., Dodson, R.J., Haft, D.H., Hickey, E.K., Peterson, J.D., Umayam, L., *et al.* (2000). DNA sequence of both chromosomes of the cholera pathogen *Vibrio cholerae*. Nature *406*, 477–483.

Holden, M.T., Titball, R.W., Peacock, S.J., Cerdeno-Tarraga, A.M., Atkins, T., Crossman, L.C., Pitt, T., Churcher, C., Mungall, K., Bentley, S.D., *et al.* (2004). Genomic plasticity of the causative agent of melioidosis, *Burkholderia pseudomallei*. Proc. Natl. Acad. Sci. U.S.A. *101*, 14240–14245.

Hubber, A., Vergunst, A.C., Sullivan, J.T., Hooykaas, P.J., and Ronson, C.W. (2004). Symbiotic phenotypes and translocated effector proteins of the *Mesorhizobium loti* strain R7A VirB/D4 type IV secretion system. Mol. Microbiol. *54*, 561–574.

Jain, R., Rivera, M.C., and Lake, J.A. (1999). Horizontal gene transfer among genomes: the complexity hypothesis. Proc. Natl. Acad. Sci. U.S.A. *96*, 3801–3806.

Jani, A.J., and Cotter, P.A. (2010). Type VI secretion: not just for pathogenesis anymore. Cell Host Microbe *8*, 2–6.

Kado, C.I. (1994). Promiscuous DNA transfer system of *Agrobacterium tumefaciens*: role of the virB operon in sex pilus assembly and synthesis. Mol. Microbiol. *12*, 17–22.

Kampfer, P., Scholz, H.C., Huber, B., Falsen, E., and Busse, H.J. (2007). *Ochrobactrum haematophilum* sp. nov. and *Ochrobactrum pseudogrignonense* sp. nov., isolated from human clinical specimens. Int. J. Syst. Evol. Microbiol. *57*, 2513–2518.

Kaneko, T., Minamisawa, K., Isawa, T., Nakatsukasa, H., Mitsui, H., Kawaharada, Y., Nakamura, Y., Watanabe, A., Kawashima, K., Ono, A., *et al.* (2010). Complete genomic structure of the cultivated rice endophyte *Azospirillum* sp. B510. DNA Res *17*, 37–50.

Lang, A.S., and Beatty, J.T. (2007). Importance of widespread gene transfer agent genes in alpha-proteobacteria. Trends Microbiol. *15*, 54–62.

Leiman, P.G., Basler, M., Ramagopal, U.A., Bonanno, J.B., Sauder, J.M., Pukatzki, S., Burley, S.K., Almo, S.C., and Mekalanos, J.J. (2009). Type VI secretion apparatus and phage tail-associated protein complexes share a common evolutionary origin. Proc. Natl. Acad. Sci. U.S.A. *106*, 4154–4159.

Leung, M.M., Florizone, S.M., Taylor, T.A., Lang, A.S., and Beatty, J.T. (2010). The gene transfer agent of *Rhodobacter capsulatus*. Adv. Exp. Med. Biol. *675*, 253–264.

Lidstrom, M. (2006). Aerobic methylotrophic prokaryotes. In The Prokaryotes,Vol 2: Ecophysiology and Biochemistry, Falkow, S., Rosenberg, E., Schleifer, K.-H., Stackebrandt, E., Dworkin, M., eds (Springer-Verlag, New York), pp. 618–634.

Lindroos, H., Vinnere, O., Mira, A., Repsilber, D., Naslund, K., and Andersson, S.G. (2006). Genome rearrangements, deletions, and amplifications in the natural population of *Bartonella henselae*. J. Bacteriol. *188*, 7426–7439.

Ma, W., Dong, F.F., Stavrinides, J., and Guttman, D.S. (2006). Type III effector diversification via both pathoadaptation and horizontal transfer in response to a coevolutionary arms race. PLoS Genet. *2*, e209.

McCann, H.C., and Guttman, D.S. (2008). Evolution of the type III secretion system and its effectors in plant–microbe interactions. New Phytol. *177*, 33–47.

McDaniel, L.D., Young, E., Delaney, J., Ruhnau, F., Ritchie, K.B., and Paul, J.H. (2010). High frequency of horizontal gene transfer in the oceans. Science *330*, 50.

MacLean, A.M., Finan, T.M., and Sadowsky, M.J. (2007). Genomes of the symbiotic nitrogen-fixing bacteria of legumes. Plant Physiol. *144*, 615–622.

Madhaiyan, M., Poonguzhali, S., Senthilkumar, M., Sundaram, S., and Sa, T. (2009). Nodulation and plant-growth promotion by methylotrophic bacteria isolated from tropical legumes. Microbiol. Res. *164*, 114–120.

Maggi, R.G., and Breitschwerdt, E.B. (2005). Isolation of bacteriophages from *Bartonella vinsonii* subsp. *berkhoffii* and the characterization of Pap31 gene sequences from bacterial and phage DNA. J. Mol. Microbiol. Biotechnol. *9*, 44–51.

Marchetti, M., Capela, D., Glew, M., Cruveiller, S., Chane-Woon-Ming, B., Gris, C., Timmers, T., Poinsot, V., Gilbert, L.B., Heeb, P., *et al.* (2010). Experimental evolution of a plant pathogen into a legume symbiont. PLoS Biol. *8*, e1000280.

Martinez-Romero, E. (2009). Coevolution in Rhizobium–legume symbiosis? DNA Cell Biol. *28*, 361–370.

Masson-Boivin, C., Giraud, E., Perret, X., and Batut, J. (2009). Establishing nitrogen-fixing symbiosis with legumes: how many rhizobium recipes? Trends Microbiol. *17*, 458–466.

Meyer, T.F., and Rosenshine, I. (2008). Host–microbe interactions: bacteria. Curr. Opin. Microbiol. *11*, 1–2.

Moulin, L., Munive, A., Dreyfus, B., and Boivin-Masson, C. (2001). Nodulation of legumes by members of the beta-subclass of Proteobacteria. Nature *411*, 948–950.

Ngom, A., Nakagawa, Y., Sawada, H., Tsukahara, J., Wakabayashi, S., Uchiumi, T., Nuntagij, A., Kotepong, S., Suzuki, A., Higashi, S., *et al.* (2004). A novel symbiotic nitrogen-fixing member of the *Ochrobactrum* clade isolated from root nodules of *Acacia mangium*. J. Gen. Appl. Microbiol. *50*, 17–27.

Nystedt, B., Frank, A.C., Thollesson, M., and Andersson, S.G. (2008). Diversifying selection and concerted evolution of a type IV secretion system in *Bartonella*. Mol. Biol. Evol. *25*, 287–300.

Oliver, K.M., Degnan, P.H., Burke, G.R., and Moran, N.A. (2010). Facultative symbionts in aphids and the horizontal transfer of ecologically important traits. Annu. Rev. Entomol. *55*, 247–266.

Paziewska, A., Harris, P.D., Zwolinska, L., Bajer, A., and Sinski, E. (2011). Recombination within and between species of the alpha proteobacterium *Bartonella* infecting rodents. Microb. Ecol. *61*, 134–145.

Pirttilä, A.M., Laukkanen, H., Pospiech, H., Myllyla, R., and Hohtola, A. (2000). Detection of intracellular bacteria in the buds of Scotch pine (*Pinus sylvestris* L.) by in situ hybridization. Appl. Environ. Microbiol. *66*, 3073–3077.

Ramsay, J.P., Sullivan, J.T., Stuart, G.S., Lamont, I.L., and Ronson, C.W. (2006). Excision and transfer of the *Mesorhizobium loti* R7A symbiosis island requires an integrase IntS, a novel recombination directionality factor RdfS, and a putative relaxase RlxS. Mol. Microbiol. *62*, 723–734.

Rankin, D.J., Rocha, E.P., and Brown, S.P. (2011). What traits are carried on mobile genetic elements, and why? Heredity *106*, 1–10.

Raymond, J., Siefert, J.L., Staples, C.R., and Blankenship, R.E. (2004). The natural history of nitrogen fixation. Mol. Biol. Evol. *21*, 541–554.

Regnery, R.L., Anderson, B.E., Clarridge, J.E., 3rd, Rodriguez-Barradas, M.C., Jones, D.C., and Carr, J.H. (1992). Characterization of a novel *Rochalimaea* species, *R. henselae* sp. nov., isolated from blood of a febrile, human immunodeficiency virus-positive patient. J. Clin. Microbiol. *30*, 265–274.

Rhomberg, T.A., Truttmann, M.C., Guye, P., Ellner, Y., and Dehio, C. (2009). A translocated protein of *Bartonella* henselae interferes with endocytic uptake of individual bacteria and triggers uptake of large bacterial aggregates via the invasome. Cell Microbiol *11*, 927–945.

Saenz, H.L., Engel, P., Stoeckli, M.C., Lanz, C., Raddatz, G., Vayssier-Taussat, M., Birtles, R., Schuster, S.C., and Dehio, C. (2007). Genomic analysis of *Bartonella* identifies type IV secretion systems as host adaptability factors. Nat. Genet. *39*, 1469–1476.

Schmid, M.C., Scheidegger, F., Dehio, M., Balmelle-Devaux, N., Schulein, R., Guye, P., Chennakesava, C.S., Biedermann, B., and Dehio, C. (2006). A translocated bacterial protein protects vascular endothelial cells from apoptosis. PLoS Pathog. *2*, e115.

Seubert, A., Hiestand, R., de la Cruz, F., and Dehio, C. (2003). A bacterial conjugation machinery recruited for pathogenesis. Mol. Microbiol. *49*, 1253–1266.

Shamaei-Tousi, A., Cahill, R., and Frankel, G. (2004). Interaction between protein subunits of the type IV secretion system of *Bartonella henselae*. J. Bacteriol. *186*, 4796–4801.

Sharp, P.M., Shields, D.C., Wolfe, K.H., and Li, W.H. (1989). Chromosomal location and evolutionary rate variation in enterobacterial genes. Science *246*, 808–810.

Stavrinides, J., McCann, H.C., and Guttman, D.S. (2008). Host–pathogen interplay and the evolution of bacterial effectors. Cell. Microbiol. *10*, 285–292.

Sullivan, J.T., Patrick, H.N., Lowther, W.L., Scott, D.B., and Ronson, C.W. (1995). Nodulating strains of Rhizobium loti arise through chromosomal symbiotic gene transfer in the environment. Proc. Natl. Acad. Sci. U.S.A. *92*, 8985–8989.

Sullivan, J.T., and Ronson, C.W. (1998). Evolution of rhizobia by acquisition of a 500-kb symbiosis island that integrates into a phe-tRNA gene. Proc. Natl. Acad. Sci. U.S.A. *95*, 5145–5149.

Teyssier, C., Marchandin, H., Masnou, A., Jeannot, J.L., de Buochberg, M.S., and Jumas-Bilak, E. (2005). Pulsed-field gel electrophoresis to study the diversity of whole-genome organization in the genus *Ochrobactrum*. Electrophoresis *26*, 2898–2907.

Toft, C., and Andersson, S.G. (2010). Evolutionary microbial genomics: insights into bacterial host adaptation. Nat. Rev. Genet. *11*, 465–475.

Tseng, T.T., Tyler, B.M., and Setubal, J.C. (2009). Protein secretion systems in bacterial–host associations, and their description in the gene ontology. BMC Microbiol. *9(Suppl. 1)*, S2.

Umemori, E., Sasaki, Y., Amano, K., and Amano, Y. (1992). A phage in *Bartonella bacilliformis*. Microbiol. Immunol. *36*, 731–736.

Vayssier-Taussat, M., Le Rhun, D., Deng, H.K., Biville, F., Cescau, S., Danchin, A., Marignac, G., Lenaour, E., Boulouis, H.J., Mavris, M., *et al.* (2010). The Trw type IV secretion system of *Bartonella* mediates host-specific adhesion to erythrocytes. PLoS Pathog. *6*, e1000946.

Wang, H.Y., Tang, H., Shen, C.K., and Wu, C.I. (2003). Rapidly evolving genes in human. I. The glycophorins and their possible role in evading malaria parasites. Mol. Biol. Evol. *20*, 1795–1804.

West, S.A., Kiers, E.T., Simms, E.L., and Denison, R.F. (2002). Sanctions and mutualism stability: why do rhizobia fix nitrogen? Proc. Biol. Sci. *269*, 685–694.

Barriers to Horizontal Gene Transfer: Fuzzy and Evolvable Boundaries

4

Fernando González-Candelas and M. Pilar Francino

Abstract

The existence of numerous types of barriers to horizontal gene transfer (HGT) is well documented. Nevertheless, no barrier is impervious, and all kinds of genes can occasionally find their way into organisms different from the ones in which they evolved. This by no means implies a free flow of genes across taxa, but, rather, the existence of complex networks of loopholes across barriers that could potentially connect all kinds of organisms through gene transfer. Loopholes may be provided by way of a small fraction of the individuals in an otherwise inaccessible population and, if exogenous genes are incorporated into the chromosome, homology-assisted recombination processes may further spread them across wild type individuals. HGT networks should be very fluid, as they depend heavily on fortuitous events and transient circumstances, such as the presence of mobile genetic elements (MGEs) in a potential donor that may extend the transfer network in particular directions. Moreover, HGT networks should be highly evolvable, as a result of (1) the multiplicity of processes with the potential to modulate, modify or alleviate the different barriers to transfer, (2) the constantly changing selective pressures operating on them and (3) the enormous plasticity of MGEs. Finally, HGT networks should be gene-specific, as different genes should have an unequal likelihood of passing through or being incorporated into new genomes.

Introduction

The controversy surrounding HGT is often polarized into views that attempt to summarize the overall impact of this phenomenon on prokaryotic evolution, particularly regarding the existence of prokaryotic species and of organismal phylogenies (Bapteste and Boucher, 2009; Doolittle and Zhaxybayeva, 2009; Puigbo *et al.*, 2009; Riley and Lizotte-Waniewski, 2009). HGT, however, encompasses a multiplicity of processes, which operate through different agents and mechanisms, rely on different ecological circumstances and result in a wide range of fitness effects on the recipient organisms. In this chapter, we wish to explore the barriers that limit HGT among organisms and the circumstances and manners in which they may be circumvented, following the rationale that taking into account the conditions in which HGT operates should contribute to the development of more specific lines of research on the detailed dynamics of HGT processes and their repercussions on the evolution of prokaryotes.

Ecological opportunity

HGT is a widespread phenomenon for many organisms. Cases have been documented involving all possible pair-wise combinations and directions of transfer in the three domains of life plus viruses (Nelson *et al.*, 1999; Boucher *et al.*, 2001; Beiko *et al.*, 2005; Choi and Kim, 2007; Keeling and Palmer, 2008; La Scola *et al.*, 2008; Hecht *et al.*, 2010; Dunning Hotopp, 2011). This may conduce to thinking that HGT is frequent at all biologically relevant levels and time-scales. Given the number of individual organisms inhabiting our planet and the time elapsed since cellular life became established on Earth, a high frequency of HGT would certainly lead to erasing most, if not all, signs of vertical inheritance. Several authors support this conclusion (Hilario and Gogarten, 1993; Glansdorff *et al.*, 2008; Koonin and Wolf, 2009) but, by large, HGT has not completely blurred the phylogenetic signal resulting from vertical transmission of genetic material. In consequence, we need to understand how different factors and processes enable or limit horizontal gene transfer among organisms.

The first step for an HGT event is the contact of the foreign genetic material and the recipient organism. We will consider the different ways in which such contacts may happen in a different section and will start here by dealing with their ecological context.

Different environments determine which assemblages can donate/receive genes from other members thus creating special groups or exchange communities. Also, the type of genes laterally transferred among members of such communities might be different because of the disparate selective advantages resulting from the transfer of specific genes in different environments. Both the size and composition of the gene pool available to be incorporated into the genome of an organism by HGT, also known as its 'supergenome' (Norman *et al.*, 2009), are dependent on the environment it inhabits. The environment not only determines the selective framework but also the set of genetic innovations accessible to cope with it, and HGT is the most common mechanism for their acquisition (Pallen and Wren, 2007). For instance, according to Pál *et al.* (2005), most changes to the metabolic network of *Escherichia coli* in the past 100 million years have been due to HGT, with little contribution from gene duplication. Furthermore, genes incorporated by HGT are mainly involved in the transport and catalysis of external nutrients, thus driving the adaptations to changing environments.

The preferential exchange of genetic material among species inhabiting the same environments may lead to the emergence of assemblages of species or communities which become bound together not only through adaptation to similar conditions but also through access and sharing of an almost common gene pool. This view has been favoured by some large scale analyses of lateral transfer which have identified preferential routes or 'highways' of gene sharing between taxa with common phylogenetic history and environmental distribution (Beiko *et al.*, 2005), even leading to the proposal of separate, independent 'genetic worlds' (Halary *et al.*, 2010), defining multiple isolated, distinct gene pools.

The phylogenomic analysis of gene content and distribution of Clusters of Orthologous Genes (COGs) among bacteria classified according to lifestyle led Audic *et al.* (2007) to conclude that clustering results from vertical inheritance and lateral gene transfer, which varies in extent and nature of transferred genes among environments. This analysis focused on *Minibacterium massiliensis*, a water-living Betaproteobacteria, and allowed the authors to conclude that bacterial communities living in this habitat exhibited the highest proportion of horizontally transferred genes, especially for three functional categories: C-COG (energy production and conversion), D-COG (cell division and chromosome partitioning) and I-COG (lipid metabolism).

Environmental conditions may have a strong effect on the actual mechanism of HGT and, in consequence, on the likelihoods and frequencies of different processes (Norman *et al.*, 2009). For instance, Charpentier *et al.* (2011) have recently shown that environmental stress conditions affect the state of competence for transformation in *Legionella pneumophila* by means of changes in the expression of competence-related genes such as *comEA*. As a result, the chances of transformation with foreign DNA may increase in stressing environments.

However, other mechanisms seem to play a more important role than transformation in certain environments. Kenzaka *et al.* (2010) have recently evaluated the rate of gene exchange among bacteria mediated by phages in aquatic environments, estimating values as high as 2×10^{-3} per total direct count when natural bacterial communities were recipients. McDaniel *et al.* (2010) analysed the purple non-sulfur bacterium *Rhodobacter capsulatus,* which inhabits marine environments. They found that gene transfer agents (GTAs), host-encoded virus-like sequences found in the genome of most Rhodobacterales, have contributed several orders of magnitude more often to the incorporation of new genetic material into the *R. capsulatus* genome than transformation and transduction. A large analysis of almost 120,000 homologous DNA families (Halary *et al.*, 2010) has recently concluded that plasmids and other vehicles of gene exchange have been and still are mainly responsible for gene exchange among bacterial chromosomes.

Potential hot spots in the environment for recombination and HGT include biofilms, the rhizosphere, decomposing material, guts of soil animals and the interior of bacterivorous protozoa (Ragan and Beiko, 2009). Hence, it is not surprising that high rates of HGT are found in species commonly found in these habitats, with a high proportion of genes transferred from other members of the same communities. At the other end, obligate intracellular bacteria have been thought to be largely unaffected by HGT owing to their isolation from potential donors (Renesto *et al.*, 2005). Nevertheless, *Buchnera,* obligate endosymbiont bacteria of aphids, still present prophages, plasmids and transposons (Bordenstein and Reznikoff, 2005) and some evidences point towards lateral gene acquisition after starting their obligate lifestyle (Van Ham *et al.*, 2000). Similar observations have been made in other intracellular parasites such as *Rickettsia felis* (Gillespie *et al.*, 2008) or *Chlamydia trachomatis* (DeMars *et al.*, 2007).

However, some evidence indicates that there is not an absolute necessity of sharing the same environment for the lateral transfer of genes between bacterial species. For instance, although generalist species, such as *Pseudomonas aeruginosa,* have large genomes and can survive in a wide range of habitats, it is somewhat surprising that a clinical strain of this species harbours genes for the degradation of secondary metabolites of trees (Mathee *et al.*, 2008). Amoebas can act as a common ground for the exchange of genes among the tens of bacterial species that inhabit, regularly or occasionally, their cytoplasm (Coscollá *et al.*, 2011), and viruses are capable of moving between habitats (Breitbart and Rohwer, 2005). In consequence, the environment plays a major role in the frequency, nature, and mechanism of horizontal gene transfer but exchange communities (Jain *et al.*, 2003) cannot be considered as closed.

Host range of different HGT mechanisms

Numerous bacterial species, belonging to very different phylogenetic and trophic groups, are naturally competent for transformation by free exogenous DNA (Lorenz and Wackernagel,

1994). In addition, an enormous variety of MGEs contributes to the spread of genes across bacterial cells. MGEs can vary widely in terms of the range of organisms that can serve as their hosts and among which they can transfer genetic material. We will explore the taxonomic breadth across which different HGT mechanisms can operate, review some of the factors that contribute to delineate this range, and analyse how interactions among different MGEs in a host cell or bacterial community can affect it.

Transformation

Transformation is the active uptake of exogenous DNA by competent cells followed by genomic integration. Naturally transformable species have been detected in all the major taxonomic groups of Eubacteria as well as in the Archaea (Lorenz and Wackernagel, 1994; Kleter *et al.*, 2005). Unlike genes responsible for conjugation and transfection, which are MGE-encoded, the genes required to render a cell transformable are usually encoded by the bacterial chromosome, and often dispersed across different chromosomal regions, so that transformation could be considered as the genuine prokaryotic sex (Lorenz and Wackernagel, 1994).

For natural transformation to occur, DNA has to be released from donor cells and dispersed or maintained in the environment until being encountered by potential recipient cells. Numerous studies have demonstrated that extracellular DNA of multiple sources can be present at high concentrations in diverse prokaryotic habitats and persist for considerable time (Nielsen *et al.*, 2007). On the other hand, recipients need to develop competency via the highly regulated expression of specific proteins and complexes that may include sophisticated DNA uptake systems. Generally, competent bacteria can translocate chromosomal DNA fragments, plasmids or phage DNA across their membrane(s) (Thomas and Nielsen, 2005; Brigulla and Wackernagel, 2010). The actual translocation process varies among taxa, but usually involves the transfer of single DNA strands (Chen *et al.*, 2005). Natural transformation machineries have been studied in a variety of bacteria, including *Bacillus subtilis, Streptococcus pneumoniae, Haemophilus influenzae, Neisseria gonorrhoeae* and *Helicobacter pylori*. Although not exclusively, most of these DNA uptake machineries are related to type IV pili and type II secretion systems (Lorenz and Wackernagel, 1994; Averhoff and Friedrich, 2003; Chen and Dubnau, 2003). Importantly, although most uptake systems can transport DNA from diverse sources, for some of them the species of origin of the DNA can strongly affect the likelihood of its uptake. In *H. pylori*, which uptakes DNA by means of a reverse type IV secretion system (Smeets and Kusters, 2002), intraspecific DNA is preferentially uptaken even in the presence of 1000-fold excess of DNA molecules from other sources; this result was obtained with both genomic DNA and PCR products, indicating that *H. pylori* and foreign DNA are not differentiated by methylation patterns but at the level of primary DNA sequence (Levine *et al.*, 2007). Specific DNA uptake sequences have not yet been identified in this species, but are well characterized in other naturally transformable bacteria such as *Haemophilus* and *Neisseria* (Saunders *et al.*, 1999). These organisms preferentially take up DNA from their own clades, which they recognize by 9- or 10-nucleotide-long sequences that are overrepresented in their genomes (Thomas and Nielsen, 2005). In contrast, the Gram-positive *B. subtilis* and *S. pneumoniae* and the Gram-negative *Acinetobacter* sp. and *Pseudomonas stutzeri* do not recognize specific sequences and can similarly uptake homologous and heterologous DNA (Averhoff and Friedrich, 2003).

Therefore, selectivity in the transformation process can exist at the level of exogenous DNA recognition, but the degree to which it occurs is extremely variable across bacterial species.

Interestingly, recent work with *Streptococcus* has revealed that competent cells can trigger DNA release from non-competent cells belonging to the same or closely related species, via what seems to be an active mechanism to secure a homologous DNA pool for uptake and recombination (Johnsborg *et al.*, 2008). This phenomenon involves a fratricide mechanism that is co-regulated with natural transformation. Competence for natural transformation is a transient strictly regulated physiological state in *Streptococcus* and many other bacteria and can be induced by a variety of stressors. In the streptococci, it can be induced by high concentrations of a secreted pheromone termed the competence-stimulating peptide (CSP), for which there exists a variety of types specific to limited subgroups of strains termed pherogroups (Havarstein *et al.*, 1995). When pherogroup members sense a critical concentration of their specific CSP, they develop the competent state and release lysins to which they are immune but that kill neighbouring cells belonging to other pherogroups. This competence-induced cell lysis has been experimentally shown to increase the efficiency of lateral gene transfer from conspecific streptococci by 1000-fold, and from closely related species by 40-fold. Similar systems of co-induced competence and bacteriocin production by CSPs have been found in other streptococci and in *B. subtilis* (Kreth *et al.*, 2005). It is clear that such systems have the potential to give rise to arms races that have likely contributed to generate the great variety of CSP molecules and cognate receptors that exist in nature (Johnsborg *et al.*, 2008).

Mechanisms that will favour the intracellular survival, genomic integration and expression of transformed DNA from closely related organisms over more divergent ones will operate once DNA has been introduced into the recipient cell. As the different factors that contribute to this selectivity will also operate for the other mechanisms of HGT, they will be discussed in a more general context in the section 'Factors operating across HGT mechanisms'.

Transduction

Because of their prevalence, phages are likely to play an important role in gene transfer in the environment. Although it used to be thought that phages only rarely crossed species boundaries, broad-host-range (BHR) generalized transducing phages that can transfer genes between species and genera have recently been discovered at high frequency in different natural environments (Nzula *et al.*, 2000; Langley *et al.*, 2003; Sullivan *et al.*, 2003). In some cases, transduction can even occur among bacteria belonging to different classes. For instance, phage SN-T can infect species within the Gammaproteobacteria, Alphaproteobacteria and Betaproteobacteria (Jensen *et al.*, 1998; Beumer and Robinson, 2005). Moreover, the range of organisms into which a phage can inject its DNA can be wider than the actual infective host range (Jensen *et al.*, 1998; Holmfeldt *et al.*, 2007).

Bacteria possess diverse types of resistance mechanisms to avoid infection by phage. Interestingly, some of these mechanisms are actually encoded themselves by prophages or plasmids. Although they serve mainly to ensure bacterial survival, resistance mechanisms are also relevant in defining phage host range and therefore the potential for transduction among different organisms. Phages, in turn, employ numerous resistance countering, and therefore host-range expanding, adaptations (Hyman and Abedon, 2010). As a result, complex arms races can develop involving the phage, the host bacterium and the different MGEs that it might contain. In addition, bacteria are often found in biofilms, which can diminish their susceptibility to phage infection.

As for transformation, the first level of resistance to phage can occur at the bacterial cell surface, at the level of phage attachment to cognate receptor molecules. Interaction between phage and bacterium can be reduced by loss, down-regulation or structural alteration of phage receptor molecules. Phage receptors, though, often have important roles in cell biology and tampering with them may entail substantial fitness costs. Nevertheless, mutations to adsorption resistance through these processes are common (Hyman and Abedon, 2010), speaking to the high selection pressure exerted by phages. Some bacteria have evolved very specific mechanisms to turn on and off the expression of a given receptor. In *Campylobacter jejuni*, invertible genome segments bound by Mu-like prophage segments control the production of a bacteriophage receptor, flagellin. Again, this system is costly to the cell, as, in the flagellin-off stage, flagella loss reduces pathogenicity and the Mu-like prophage genome is restored and activated, leading to production of infectious phage (Scott *et al.*, 2007). Other prophages are known to interfere with the display of their own cell-surface receptors (Perry *et al.*, 2009).

Phages attach to their host receptors via specific molecules, known as antireceptors. Possibly in order to counteract the host capacity to alter receptor molecules, phage antireceptors are often very flexible and able to recognize several bacterial surface molecules (Goldberg and Murphy, 1983; Montag *et al.*, 1990; Hyman and Abedon, 2010). In some cases, phages encode multiple antireceptor genes for different receptors, displaying more than one antireceptor in the same virion (Scholl *et al.*, 2001; Leiman *et al.*, 2007) or utilizing site-specific genome inversion mechanisms to switch between antireceptors (Hyman and Abedon, 2010). Phages can further respond to a receptor mutation in their hosts by changing their antireceptors so that they recognize the altered receptor or a different one, or by becoming less stringent in terms of receptor recognition. Remarkably, diversity-generating systems specifically targeting antireceptor genes have recently been discovered, such as the reverse-transcriptase-mediated, site-specific antireceptor gene mutation system of *Bordetella* phages BPP-1, BMP-1, and BIP-1 (Liu *et al.*, 2002; Medhekar and Miller, 2007). Logically, antireceptor-modifying mutations may indirectly result in shifted or expanded phage host ranges. Relaxed host specificity can also result from conformational changes induced by modification of phage proteins other than the antireceptor molecule (Hyman and Abedon, 2010).

Rather than eliminating, downregulating or modifying phage receptor molecules, bacteria can also conceal them by means of physical barriers, such as capsules and other extracellular structures. However, such barriers can also be circumvented by the phage, via enzymes that degrade their constituent polymers. Typically, these enzymes are active against only a narrow range of barrier types (Sutherland *et al.*, 2004; Scholl and Merril, 2005) and extracellular polymers should block infections by only certain phages. Therefore, the broad chemical diversity of extracellular barriers among bacterial strains could be the product of a phage-mediated frequency-dependent selection (Abedon, 2009) where rare barrier types would be less subject to barrier-circumventing phages (van der Ley *et al.*, 1986; Scholl and Merril, 2005). Furthermore, while extracellular polymers reduce infection susceptibility to some phages, they can increase susceptibility to other phages that may recognize them as receptor molecules (Pelkonen *et al.*, 1992; Hanfling *et al.*, 1996; Weinbauer,

2004). Consequently, arms races could develop involving a variety of competing phage and bacterial strains.

After adsorption, different mechanisms can operate to avoid the delivery of the phage genome to the cytoplasm. These mechanisms can be encoded by plasmids and prophages or by the bacterial chromosome (Hyman and Abedon, 2010). If the phage genome reaches the cytoplasm, several other processes can still interfere with its integrity or its capacity to integrate into the bacterial genome. These processes are addressed below in the section 'Factors operating across HGT mechanisms'. Finally, some bacteria may respond to phage infection by employing measures that will cause the infected cell to die, presumably to decrease the probability of infection of other cells in the population. A widespread mechanism of this sort involves nucleases that, when activated, cleave the anticodon loop of a specific tRNA, blocking protein synthesis (Hyman and Abedon, 2010). Some phages, such as phage T4, however, are able to repair these cuts, while others have evolved alternative means of resistance (Amitsur *et al.*, 1987; Labrie and Moineau, 2007; Haaber *et al.*, 2009).

Conjugation

Amongst the bacterial gene exchange mechanisms, transformation, transduction and conjugation, the latter has the broadest host range of transfer (Dreiseikelmann, 1994; Droge *et al.*, 1998). Conjugation is widespread among Gram-negative and Gram-positive bacteria and occurs also in Crenarchaeota (Stedman *et al.*, 2000). It is mediated by plasmids or by integrative conjugative elements (ICEs). A conjugative plasmid recombined into the chromosome can also initiate conjugal transfer of chromosomal DNA. Conjugative elements are able to induce the formation of mating pairs between a donor and a recipient cell. Conjugative plasmids and ICEs of Gram-negative bacteria encode filamentous structures called *pili* that aid in mating-pair formation (Beaber *et al.*, 2002; Burrus *et al.*, 2002), whereas elements that are restricted to Gram-positive bacteria produce surface-presented protein factors that induce cell aggregation (Clewell, 1993; Godon *et al.*, 1994; Burrus *et al.*, 2002). In the best-studied conjugation system, the one mediated by plasmids of Gram-negative bacteria, the donor cell produces a mating pair formation (Mpf) system that enables intimate physical contacts with a recipient cell through a membrane-spanning protein complex and a surface-exposed sex pilus. Additionally, a coupling protein (CP) interacts with the DNA substrate and couples it to the secretion pore formed by the Mpf system. Mpf/CP conjugation systems belong to the family of type IV secretion systems, which also includes DNA uptake and release systems, as well as effector protein translocation systems (Schroder and Lanka, 2005). The translocated DNA is a single strand that starts from an origin of transfer (oriT) and enters the recipient cell directed by a relaxase protein covalently joined to the 5′ DNA end (Lee and Grossman, 2007). Much less is known about how DNA traverses the recipient cell membranes. Pérez-Mendoza and de la Cruz (2009) performed a systematic analysis of the contribution of each gene in an *Escherichia coli* recipient to the conjugal transfer of IncW plasmid R388. Their results indicated that no non-essential recipient genes are required or affect conjugative DNA processing or transport in a significant manner, although mutations in lipopolysaccharide synthesis pathways can impair mating pair formation, particularly in liquid matings. This screen did not reveal any molecule that might serve as a 'conjugation receptor', analogous to the receptor molecules of transformation and transfection. Therefore, conjugal DNA transfer appears to be a process that is basically carried out by the donor cell and where the recipient has little say (Pérez-Mendoza and de la Cruz, 2009). A 'shoot and pump' conjugation model (Llosa *et al.*, 2002) has been proposed where the sheer push force imparted by the type IV secretion system on the DNA causes the transport machinery to act as a syringe or a drilling machine for injecting the DNA into the recipient cell.

After DNA entry into the recipient, the requirements of plasmids and ICEs for replication and maintenance differ, and we will discuss them, and their effects on host range, separately in the next two sections.

Plasmids

Once inside a recipient cell, the transferred plasmid DNA has to be reconstituted into a replication-proficient form. In many cases, this process must involve, at least, recircularization of the transferred DNA strand and synthesis of a complementary, lagging strand. In the analysis of recipient gene contribution to conjugal transfer performed by Pérez-Mendoza and de la Cruz (2009), no specific recipient enzymes appeared to be required for the reconstitution of a replication-proficient plasmid. The IncW R388 plasmid used in their experiments encodes a relaxase that is transported to recipient cells during bacterial conjugation and is likely responsible for DNA strand recircularization (Garcillán-Barcia *et al.*, 2007). However, R388 does not encode a DNA primase for lagging strand synthesis, and it probably relies on the host RNA polymerase, an enzyme essential for cell survival, as occurs for plasmids that employ a rolling-circle mode of replication (Kramer *et al.*, 1997). In other systems, the primase is plasmid-encoded and its transport to the recipient cell is essential for conjugation in some heterologous matings (Merryweather *et al.*, 1986). For plasmids that rely on host relaxases and primases, the ability of these enzymes to be functional on plasmid DNA will determine if a functional plasmid can be reconstituted after transfer and contribute to define whether a given species can be counted within the plasmid's host range.

Probably, the single most important factor in determining plasmid host range is the capacity of sustaining replication in a particular recipient after transfer and reconstitution (Guiney, 1982; del Solar *et al.*, 1996). Plasmids that can be replicated with fewer host-dependent requirements can successfully replicate in more cytoplasms. A critical time point is the initiation of replication, as plasmids that require host proteins for this process are limited to bacteria whose proteins are able to interact productively with the plasmid's replication origin (Caspi *et al.*, 2000). In contrast, BHR plasmids often initiate replication in an autonomous manner. This is the case for the very promiscuous IncQ plasmid RSF1010, which can be transferred and replicate among Proteobacteria, Firmicutes, Actinobacteria and Cyanobacteria, and even reach eukaryotic organisms including yeasts, plants, and animals (Gormley and Davies, 1991; Sode *et al.*, 1992; Nesvera *et al.*, 1994; Rawlings and Tietze, 2001). IncQ plasmids do not encode their own conjugative machinery, but they can be mobilized by a variety of conjugative elements and transmitted by means of different type IV secretion systems, which also contributes greatly to their very broad host ranges. They replicate by a strand-displacement mechanism and contribute to their replication by means of three proteins: RepC, which binds specifically to an *oriV* origin of replication; RepA, a 5′ to 3′ DNA helicase; and RepB, a DNA primase (Scherzinger *et al.*, 1991). Comparative analyses using cell extracts of *E. coli* and *P. aeruginosa* strongly suggest that IncQ plasmids employ the same mechanism of replication in different hosts (Diaz and Staudenbauer, 1982).

Plasmids that do rely on host proteins for initiating replication can nevertheless have very broad host ranges, as long as they exhibit substantial flexibility in terms of their interaction requirements (del Solar *et al.*, 1996; Doran *et al.*, 1999). This is the case of plasmids belonging to the IncP-1 group, which are among the most promiscuous plasmids known so far. IncP-1 plasmids have the ability to transfer between and replicate in nearly all species of the Alpha-, Beta- and Gammaproteobacteria, and, for the IncP-1 plasmid pKJK10, transfer has also been detected between Proteobacteria and Actinobacteria cohabiting the rhizosphere of barley plants (Musovic *et al.*, 2006). IncP-1 plasmids employ a θ replication mode that starts at an *oriV* replication origin and encode the initiation protein TrfA. The initiation of IncP-1 plasmid replication also requires several host proteins, including DnaA, DnaB and DnaC (Pinkney *et al.*, 1988). DnaA modulates the action of TrfA by binding to DnaA boxes and stabilizing the formation of the TrfA-mediated open replication–initiation complex. Genetic information indicates that, at least in the IncP-1α subgroup, host promiscuity is partially linked to a modular organization of *oriV*, as this origin contains several DnaA boxes that are differentially bound by DnaA in different hosts (del Solar *et al.*, 1996). Bound DnaA interacts with DnaB helicase to form a DnaABC complex involved in priming. The process of DnaB loading and activation at *oriV* may also be different in various hosts (Caspi *et al.*, 2000). Moreover, the observation that DnaA boxes are not required for the replication of IncP-1α plasmids in *P. aeruginosa* raises the possibility of an alternative mechanism for the stimulation of open complex formation and DnaB loading at *oriV* (Doran *et al.*, 1999).

In contraposition to BHR plasmids, which are able to transfer across different phyla and even across the prokaryote/eukaryote barrier, those that have more restricted transfer capacities are termed NHR (narrow host range) plasmids. Some NHR plasmids can transfer across family-level groups of bacterial hosts, whereas others are apparently restricted within a single bacterial genus. However, selection experiments have been successful at enlarging the host-range of genus-specific plasmids via, for instance, mutations in the plasmid gene encoding the initiator of replication protein (Garcia de Viedma *et al.*, 1996). Thus, minor genetic changes in a plasmid replication protein can expand the plasmid host range (del Solar *et al.*, 1996). Nevertheless, other factors may restrict host range enlargement in nature, as phylogenetic analyses have shown that NHR plasmids can have tight, long-term evolutionary associations with specific host genera. This is the case of the RepA_N family of plasmids, which is broadly distributed in Gram-positive bacteria but where each member of the family has a narrow, intrageneric, host range. Congruence between bacterial host phylogenies and those based on the replication-initiator protein RepA indicates that this plasmid family predated the divergence of the Firmicutes and that different plasmids coevolved with specific genera. Other RepA_N plasmid loci, however, have undergone different phylogenetic histories, indicating a modular evolution for this plasmid's genome (Weaver *et al.*, 2009).

The long-term maintenance of a plasmid in a given host necessitates other conditions besides effective replication. For instance, the BHR IncQ plasmids also ensure their host range by virtue of a partitioning system that is active in a number of different hosts (Siddique and Figurski, 2002) and, in some groups of plasmids, such as IncQ2, by means of toxin-antidote systems that contribute to plasmid stability when copy-number is low (Deane and Rawlings, 2004). These mechanisms, added to their capacity of being mobilized by a variety of conjugative elements and to their autonomous initiation of replication, ensure that IncQ plasmids not only have one of the broadest host ranges of all MGEs, but may just be amongst the most successful plasmids in nature in terms of their prevalence in a variety of environments (Meyer, 2009). Another interesting case is that of IncP-1 plasmids. Although these plasmids can replicate in numerous different hosts, they are not maintained with equal efficiency in each of them. Importantly, plasmid stability does not correlate with the phylogenetic relatedness of the hosts, but rather seems to be strain specific, varying greatly within the same genus or even the same species. This implies that plasmid stability can be affected by rapidly evolving host-encoded factors (De Gelder *et al.*, 2007; Heuer *et al.*, 2007). On the other hand, experimental evolution studies have shown that IncP-1 plasmids are able to adapt to hosts in which they were initially unstable, and thereby expand their long-term host range (De Gelder *et al.*, 2008). In nature, adaptation to new hosts is likely to be aided by recombination among different IncP-1 plasmids, which has been shown to be a prominent feature of their evolution (Norberg *et al.*, 2011). Interestingly, the actual host range of IncP-1 plasmids in bacterial communities can also be affected by the identity of the bacteria *donating* the plasmid and by environmental conditions (De Gelder *et al.*, 2005). Given all of the factors involved in defining their host range, IncP-1 transfer networks, and those of many other plasmids, could be highly non-random in nature and vary greatly among different environments.

Integrative conjugative elements

ICEs are a diverse group of genetic elements that encode the functions required to integrate into their bacterial host chromosome and to transfer between cells by conjugation. Upon induction, ICEs excise from the chromosome by site-specific recombination, transfer the resulting circular form by conjugation and integrate by recombination between a specific site of this circular form and a site in the genome of their host. Although their replication and stability is basically ensured by way of their integration into the host genome, some ICEs have recently been shown to be also able to replicate in an autonomous manner after excision; this step does not seem to be required for mating, but apparently facilitates reinsertion of the element into the donor chromosome by increasing its copy number and maintaining it through several generations (Lee *et al.*, 2011). The ICEs include conjugative transposons and conjugative transposon-like genomic islands, as well as numerous unclassified MGEs. They can be found both in the chromosome and in plasmids and they show enormous variation in their genetic organization, the genes responsible for their insertion and excision and the accessory genes they carry. The degree of target site specificity of ICE insertion reactions also varies greatly, even for elements that use closely related enzymes (Roberts *et al.*, 2008). Some elements can also use specific sites in one host and multiple sites in a different host if the preferred site is not present (Wang *et al.*, 2000), as has been shown for Tn*916* in different strains of *Clostridium difficile* (Hussain *et al.*, 2005).

Genome sequence analyses suggest that ICEs are widespread in bacteria. Therefore, like phages and plasmids, they probably contribute importantly to horizontal gene transfer. Also similarly to other MGEs, host range varies widely among different ICEs. Some elements transfer only between conspecific bacterial strains, whereas others transfer between different species, genera or phyla (Clewell *et al.*, 1995; Scott and Churchward, 1995; Burrus *et al.*, 2002). For example, the conjugative transposon Tn*916* has a very broad host range and can transfer from *Enterococcus faecalis* to a large number of Firmicutes and to various Proteobacteria (Clewell *et al.*, 1995).

ICEs show a highly modular gene organization and mode of evolution and are able to exchange gene modules among themselves as well as with plasmids, phages and non-conjugative transposons. Promiscuous module exchange probably contributes significantly to the evolution of both integration and transfer specificity in ICEs and, as the integration and conjugation functions are encoded by different modules, these two properties can probably evolve and affect host range quite independently of each other (Burrus *et al.*, 2002).

Factors operating across HGT mechanisms

Restriction-modification systems

DNA restriction will constitute one of the first barriers to the survival of exogenous DNA within the cell. Restriction–modification (R–M) systems are ubiquitous in eubacteria and archaea. Most R–M systems comprise DNA methyltransferase and restriction endonuclease activities. The methyltransferase enables recognition of 'self' DNA by methylation of specific nucleotides within particular DNA sequences, whereas the endonuclease enzymatically cleaves foreign unmodified DNA. There are thousands of different R–M systems, which exhibit extensive diversity in their DNA sequence recognition specificities (Bayliss *et al.*, 2006). In the common type II R–M systems, specificity is conferred by an autonomous target recognition domain (TRD), which is present in both the methyltransferase and the endonuclease, enabling independent binding of each enzyme to the recognition site (Bujnicki, 2001). Type I systems contain a single specificity subunit that has two DNA-binding domains and is present in both enzymes. The gene encoding this subunit exhibits high levels of divergence within individual species and can undergo recombinatorial reassortment of the two DNA-binding domains resulting in enzymes with altered recognition sites (Dybvig *et al.*, 1998; Murray, 2000). Type III systems are less well characterized, but phylogenetic analyses indicate that regions likely to be involved in sequence specificity are under positive selection for variability and undergo intergeneric HGT (Bayliss *et al.*, 2006).

An efficient R–M system can drastically reduce the frequency of interspecies and even intraspecies HGT (Tock and Dryden, 2005; Hoskisson and Smith, 2007). However, several HGT mechanisms transfer DNA in a single-stranded form and ssDNA is thought to be rather refractory to the action of R–M systems (Thomas and Nielsen, 2005). In the case of phage or plasmid DNA that enters in a single-stranded form by transduction or conjugation, DNA becomes susceptible to restriction after synthesis of the second strand. Transformation is therefore the mechanism that is the least susceptible to restriction. For instance, *Streptococcus pneumoniae* contains R–M systems that restrict transduction among strains by factors of 10^{-5}–10^{-6}, whereas they only restrict plasmid transformation by half and they do not affect at all transformation with chromosomal DNA (Lacks and Springhorn, 1984). These different susceptibilities can be attributed to the fact that phage DNA is introduced into the cell in duplex form whereas transforming DNA enters as a single strand. In the case of transformed plasmids, complementing strand fragments from separately entering molecules form partially double-stranded intermediates that are then susceptible to cleavage by restriction enzymes (Saunders and Guild, 1981). Nevertheless, R–M systems in other organisms have been shown to affect transfer by transformation. Several restriction enzymes are known to mediate cleavage of ssDNA (Nishigaki *et al.*, 1985), and restriction of ssDNA has been shown to be a barrier to natural transformation in *Pseudomonas stutzeri* (Berndt *et al.*, 2003). Moreover, dsDNA has been detected after transformation of pathogenic *Neisseria* species (Jyssum *et al.*, 1971; Biswas and Sparling, 1981) and it has been proposed that this genus may undergo transformation both by ssDNA, which could be used directly by RecA to mediate homologous recombination, and by dsDNA, which would require the DNA helicase activity of the RecBCD enzyme and would be susceptible to restriction (Taylor and Smith, 1980; Mehr and Seifert, 1998). Restriction is known to affect the population structure of *N. meningitidis*, which clusters into clades that are associated with different R–M systems, and transformation has been implicated in the generation of this structure because the density of DNA uptake sequences in different regions of the genome correlates strongly with the recombination rate (Budroni *et al.*, 2011). The capacity of restriction to limit intra- and interspecific transformation has also been demonstrated in *Staphylococcus aureus*, which contains several types of R–M systems that differ among isolates (Corvaglia *et al.*, 2010).

Interestingly, in spite of the protection that R–M can offer against infectious agents, several factors suggest that selective pressures to maintain these systems are sometimes alleviated. Many R–M systems display mechanisms that enable phase variation (PV), i.e. reversible, high frequency changes between on and off gene expression (Bayliss *et al.*, 2006) and R–M mutants that are hypersusceptible to the transfer of DNA from foreign sources are often found in nature (Sung and Lindsay, 2007; Corvaglia *et al.*, 2010). In strains of *Helicobacter pylori*, a type II R–M system is flanked by 80-bp direct repeats that facilitate its spontaneous deletion, but the system is easily reacquired by natural transformation (Aras *et al.*, 2001). In addition, many R–M systems are encoded in plasmids and can therefore also be easily lost and reacquired. These scenarios likely reflect the wide array of fitness effects potentially caused by foreign DNA from different sources or under different circumstances.

On the other hand, selective pressures clearly favour escape from R–M systems for MGEs. In plasmids and phage, mutation and selection frequently favour the loss of restriction sites or their disposition in arrangements that are less deleterious. For instance, the large genome of phage T7 is insensitive to the type III R–M system of *S. aureus*, because, even though it contains 36 recognition sites for this system's endonuclease, they are all present in the same orientation, whereas efficient restriction requires the head-to-head orientation of two consecutive sites for cleavage to occur (Corvaglia *et al.*, 2010). The selection pressure to loose recognition sites in MGEs likely results in a counteracting pressure for the evolution of new target recognition specificities in R–M systems, contributing to their large variability. Beyond recognition sequence loss and rearrangement, many phage and conjugative plasmids have specific mechanisms to evade the effects of host restriction systems, including the masking of restriction sites and the inactivation or modulation of R–M enzymes. For instance, in T-even phages DNA glucosylation is employed as a means of protection from most R–M systems, while T3 and T7 produce anti-restriction proteins that bind to type I restriction complexes and prevent them from binding to DNA (Murray, 2002). Interestingly, the structure of one of these proteins suggests that it neutralizes R–M complexes by mimicking the shape of the DNA substrate (Bandyopadhyay *et al.*, 1985; Atanasiu *et al.*, 2001). DNA mimicry has been shown to operate in other anti-restriction systems, such as the Ard (alleviation of restriction of DNA) proteins of conjugative plasmids (McMahon *et al.*, 2009). The proteins required for anti-restriction functions are often co-transferred with the MGE's DNA or produced early, so that they can anticipate the action of restriction systems. In the case of conjugative plasmids, this may entail transcription from special promoters recognized within secondary structures of the single DNA strand that enters the host cell (Bates *et al.*, 1999), allowing for the accumulation of anti-restriction proteins before the transferred strand is converted into duplex DNA. Finally, conjugative plasmids can also survive restriction by transiently saturating the system through the transfer of multiple plasmid copies or of extensive regions of chromosomal DNA (Matic *et al.*, 1996).

In addition to their protective role against infection, R–M systems may also have other relevant consequences for HGT, as restriction endonucleases generate recombinogenic DNA ends and could therefore potentiate the recombination of DNA fragments into the host chromosome (Matic *et al.*, 1996; Murray, 2002). Moreover, the fragmentation of incoming DNA may serve to separate potentially advantageous coding sequences from neighbouring deleterious ones (Milkman *et al.*, 1999).

CRISPR-based systems

A sophisticated strategy of defence against phages and plasmids has been recently uncovered in prokaryotes. This defence system features a unique type of repetitive DNA stretches, termed 'clustered regularly interspaced short palindromic repeats' (CRISPRs). CRISPRs consist of identical repeats of DNA sequences separated by highly variable spacer sequences of phage or plasmid origin (Mojica *et al.*, 2005). Small RNAs produced from the spacer sequences guide CRISPR-associated (Cas) proteins to recognize and degrade, or otherwise silence, invading MGEs containing complementary sequences. In this way, CRISPR-Cas systems function as a sort of adaptive immunity that allows efficient recognition and neutralization of previously encountered foreign elements (Barrangou *et al.*, 2007; Marraffini and Sontheimer, 2008). There are multiple variations of CRISPR systems found in both Eubacteria and Archaea, mediated by largely distinct components and mechanisms (Terns and Terns, 2011). These systems are encoded by operons that have an extraordinarily diverse architecture and a high rate of evolution for both the *cas* genes and the unique spacer content. Classification systems are emerging to organize the different CRISPR-based systems of prokaryotes with basis on their sequence organization and evolution (Makarova *et al.*, 2011). In addition, the high variability of the CRISPR loci can be exploited for phylogenetic and evolutionary studies of closely related species and strains (Liu *et al.*, 2011) and for understanding the short-term dynamics of exposure to phages in bacterial populations and communities (Pride *et al.*, 2012). Moreover, given the importance of phage and plasmids in bacterial niche specialization, differences in CRISPR loci can be heavily involved in delineating intraspecific population structure and adaptive sublineage evolution (Fricke *et al.*, 2011).

Many questions still remain about the mechanisms of action of CRISPR systems and their regulation, but it is clear that these systems operate under tight regulatory control. In enteric bacteria, the expression of CRISPR arrays and associated *cas* operons is repressed by the histone-like nucleoid structuring protein (H-NS; Pul *et al.*, 2010), a well-known dual regulator of gene expression in Gram-negative bacteria. H-NS-mediated silencing can be counteracted by the transcription factor LeuO (Westra *et al.*, 2010). However, expression of *leuO* and *cas* genes is not induced by phage infection, suggesting that additional factors are involved in H-NS derepression. It is possible that incoming genetic elements compete with CRISPR/*cas* promoters for H-NS binding (Pul *et al.*, 2010; Westra *et al.*, 2010), given that H-NS has preference for binding AT-rich DNA and that horizontally transferred elements are normally enriched in ATs (Rocha and Danchin, 2002). Interestingly, H-NS generally binds and represses foreign DNA through a process known as xenogeneic silencing (Navarre *et al.*, 2006, 2007; Oshima *et al.*, 2006), and therefore can be thought of in itself as a self-protection mechanism, whose titration by large numbers of foreign AT-rich molecules within the cell would represent an alert signal that would recruit the CRISPR systems of defence (Mojica and Díez-Villaseñor, 2010).

Specific mechanisms through which phages may achieve resistance to CRISPR systems have not yet been described, but the high rates of sequence evolution by mutation or recombination undergone by many phages will facilitate their escape from memory-based immunity. Metagenomic analyses reveal that phage mosaicism in natural communities can be very high, involving mosaic fragment sizes smaller than the average CRISPR spacer, such that some phages in a given population may evade CRISPR blockage and continue to productively infect (Andersson and Banfield, 2008).

Recombination and DNA repair

Exogenous DNA that survives restriction and CRISPR systems may be integrated into the bacterial host chromosome, or another autonomous replicon, in order to persist in the recipient cell. By far, the most efficient mechanism of integration is homologous recombination (HR), but this mechanism is constrained by DNA sequence divergence. It has been demonstrated in various taxonomic groups that the efficiency of HR decreases with sequence divergence between host and incoming DNA (Shen and Huang, 1986; Majewski *et al.*, 2000; Meier and Wackernagel, 2005). Divergence affects the activity of recombination enzymes, particularly at the initial stage of the strand exchange process, which requires a minimum length of homology or minimum efficient processing segment (MEPS), below which recombination becomes inefficient (Shen and Huang, 1986). Divergence reduces recombination by reducing the number of MEPSs available for initiation (Shen and Huang, 1989). If initiated, strand exchange could proceed in spite of large numbers of nucleotide differences and even large insertions (Bianchi and Radding, 1983), but the methyl-directed mismatch repair system (MMRS) would detect the base-pair mismatches and unpaired bases in the resulting heteroduplex molecule and abort the recombination process (Shen and Huang, 1986; Worth *et al.*, 1994). Recombination between diverged sequences increases significantly in strains that are deficient in MMRS. For example, recombination in MMRS-deficient genetic backgrounds increases up to 10^3-fold between *Salmonella enterica* and *E. coli* (Rayssiguier *et al.*, 1989; Matic *et al.*, 1995) and 10^2- to 10^3-fold between serovars of *S. enterica* whose genomes differ only by 1–2% at the DNA sequence level (Zahrt *et al.*, 1994). Importantly, surveys of natural populations of *E. coli* have found that about 0.1–1% of cells are MMRS mutants and therefore have an enhanced chance of integrating DNA from different species into their genome by HR (Matic *et al.*, 1995). Another important factor affecting recombination is the SOS system, which actually has a facilitating effect that runs counter the inhibiting operation of the MMRS. The SOS system is a set of physiological responses that is induced by exposure of bacterial cells to a variety of conditions that damage DNA and/or interfere with its replication. The inducing signal is thought to be persistent ssDNA and, accordingly, the SOS is induced during conjugation, transformation, restriction of foreign or host DNA, transposition and whenever plasmid or phage DNA persists in single-stranded form. In fact, the SOS also appears to be induced under normal environmental conditions (Taddei *et al.*, 1995) and is required for efficient recombination during chromosomal conjugation and transformation because it stimulates overproduction of recombination enzymes. The degree of SOS induction is stronger in interspecific conjugation, probably because the difficulty in detecting MEPS to initiate recombination with divergent sequences results in longer persistence of ssDNA in the cell. Thus, the SOS system acts as an inducible positive regulator of interspecies recombination, although it does not affect the activity of the MMRS, which remains a powerful inhibitor of recombination with divergent sequences even during strong SOS induction (Matic *et al.*, 1995). Consequently, the frequency of gene exchange by HR decreases with phylogenetic distance (Shen and Huang, 1986; Majewski *et al.*, 2000; Meier and Wackernagel, 2005).

Despite MMRS action, HR can participate in the chromosomal integration of foreign, non-homologous DNA. HR can proceed fairly efficiently *in vivo* between DNAs that differ by large insertions. If a segment of foreign DNA is embedded in sequences homologous to the recipient DNA, two HR events occurring in the flanks can effectively integrate it into the chromosome (Brigulla and Wackernagel, 2010). In natural transformation with homology-embedded foreign DNA, quantitative experiments revealed that homologous flanks of several hundred nucleotides suffice for integration and that longer flanks increase the integration frequencies by up to 100-fold (Simpson *et al.*, 2007). With the help of long homologous flanks, huge foreign DNA segments of up to 55 kb were integrated during transformation of *B. subtilis* (Itaya, 1999) and foreign DNA integrated by HR has been identified in the genome of *Mycobacterium* (Krzywinska *et al.*, 2004). The integration of foreign DNA can also proceed via homology-facilitated illegitimate recombination (IR), a combination of HR and IR. This synergy requires the presence of an anchor region, i.e. a single, short stretch of homology within the foreign DNA (de Vries and Wackernagel, 2002). The anchor region gets associated to the homologous recipient DNA by HR, and this step enhances the likelihood of IR in the adjacent foreign DNA. IR events at one or both flanks of the anchor can be stimulated, so that thousands of nucleotides can be integrated with anchor segments of just a few hundred bases. The efficiency of this mechanism varies greatly among species, but is always orders of magnitude below that of HR (Brigulla and Wackernagel, 2010). Mutants defective for recombination enzymes including RecBCD, SbcCD DNase, and RecJ DNase display up to 20-fold increased frequencies of homology-facilitated IR (Harms and Wackernagel, 2008) and disruption of MMRS results in an increased net gain of foreign DNA and the appearance of hot spots for IR events (Meier and Wackernagel, 2005). Therefore, HR of homology-embedded foreign DNA and homology-facilitated IR mechanisms can promote the introgression into a species genome of sequences that have integrated into the chromosome of individual members of the species by any other means, including long genomic islands or prophages, and can be facilitated by genetic backgrounds deficient in MMRS and other enzymes involved in the process of recombination.

IR can also integrate foreign DNA into recipient chromosomes without help from homology regions or HR participation. A stretch of foreign DNA can be integrated by two IR events, although such integration is expected to be very rare considering that single IR events occur at frequencies of 10^{-6} to 10^{-10}. Nevertheless, experimental work has shown that such double IR events are detectable and the integration of 2 kb fragments of foreign DNA into the genome of *Acinetobacter* has been detected at frequencies of 10^{-13} (Hulter and Wackernagel, 2008). The generation of mosaics between host and phage DNA has also recently been shown to occur by IR events (Morris *et al.*, 2008).

In addition to HR and IR, other integration mechanisms can also operate for foreign DNA. More than a hundred different enzymes, nearly all encoded by MGEs, recombine DNA at short specific nucleotide sequences, without the long stretches of homology required for HR, by means of a process involving DNA cleavage and reunion without DNA synthesis. Site-specific recombinases act as single enzymes or in complexes with other proteins and can catalyse the integration, excision, and inversion of DNA segments. They are responsible for the insertion of temperate bacteriophages and ICEs into new locations and for the continuous exchange of genes among these elements and bacterial chromosomes (Brigulla and Wackernagel, 2010).

Barriers and fitness costs for expressing foreign genes

Genes from different species have different GC contents and codon usage biases and several aspects of gene expression are highly tuned to the particular sequence characteristics of each genome. Such species-specific tuning can represent a substantial barrier to interspecific gene exchange by compromising foreign gene expression in different ways. One of these is represented by the histone-like nucleoid-structuring protein (H-NS), which is one of the major protein components of the nucleoid structure in bacteria and binds large regions of DNA, down-regulating gene expression, with a marked preference for AT-rich sequences. In general, horizontally transferred sequences tend to be AT-rich, and, therefore they are often recognized by H-NS and transcriptionally repressed, in a process termed xenogeneic silencing (Lucchini *et al.*, 2006; Navarre *et al.*, 2006, 2007). The expression of foreign genes can also be compromised because the host RNA-polymerase and/or transcription factors may not efficiently recognize the promoters and other regulatory sequences evolved in a different genomic background. At the translational level, expression can be affected by the codon biases of foreign genes, especially for those genes that were highly expressed in their genomes of origin. Such genes utilize the most species-biased sets of codons, because they are most strongly selected for translational efficiency and thus for using the codons that correspond to the most abundant tRNAs in the cell, which vary across species (Ikemura, 1985; Sharp and Li, 1986; Kanaya *et al.*, 1999; Rocha, 2004; Tuller, 2011).

Even when a foreign gene could be effectively expressed in a novel genome, the fitness cost of expressing it might result in selection against the recipient cell. If the gene product represents no selective benefit, cells that express it will undergo unnecessary expenditures in transcription and translation. Of course, foreign gene expression could also have negative effects on the functioning of the cell, especially if the expressed proteins affected the operation of basic cell processes. In particular, proteins that are highly interactive can interfere with resident protein complexes or protein interaction networks, due to the generation of stochiometric imbalances or to suboptimal activity in comparison with the native homologues. The fitness effects ensuing from the interspecific transfer of different classes of genes were further discussed in Chapter 1.

Interactions among MGEs

Presence of a given MGE in a host cell can affect in a positive or in a negative manner the likelihood that another element is acquired or maintained. A classical example of negative interaction is phage superinfection immunity, whereby gene expression of an incoming phage is impeded by repressors encoded by a similar prophage. Among conjugative plasmids, transfer into a recipient cell containing the same or a closely related plasmid is inhibited by a mechanism known as entry or surface exclusion. All investigated conjugative plasmids contain at least one entry exclusion gene (Garcillán-Barcia and de la Cruz, 2008). This exclusion system is believed to confer an evolutionary advantage to the plasmid as it frees the plasmid from competition at segregation during cell division, and protects the plasmid-bearing cell from too many conjugation events (Thomas and Nielsen, 2005; Garcillán-Barcia and de la Cruz, 2008). Laboratory experiments suggest that surface exclusion systems can inhibit the entry of a similar plasmid by up to 500-fold, as in the case of the R388 plasmid discussed above (Garcillán-Barcia and de la Cruz, 2008; Pérez-Mendoza and de la Cruz, 2009). Moreover, even if entry exclusion fails, the long-term coexistence of plasmids sharing replication and partitioning mechanisms is unsustainable, which serves as a basis for their classification in incompatibility (Inc) groups. Similarly, some ICEs, although not all, affect the likelihood of acquisition of closely related elements (Burrus *et al.*, 2002). Negative interactions can also occur among MGEs of very different types. For instance, the conjugative *Lactococcus* plasmid pNP40 protects cells from infection by phages via several mechanisms that include prevention of phage DNA entry into the cell as well as a R–M system (O'Driscoll *et al.*, 2006). In addition, host factors can modify the interaction between MGEs. For instance, the presence of incompatibility among similar plasmids can depend on the *E. coli* strain serving as host (Grant *et al.*, 1980).

On the other hand, the presence of different MGEs in a host cell can also enable or enhance an element's capacity to transfer, such as in the widespread phenomenon of mobilization of non-conjugative plasmids by conjugative elements. Another important synergy among MGEs derives from the fact that the ability to generate biofilms and other multicellular bacterial structures is often encoded in plasmids (Ghigo, 2001; Norman *et al.*, 2008). As biofilms are ideally suited to the exchange of genetic material of various origins, the acquisition of a biofilm-inducing plasmid in a bacterial community will affect the subsequent likelihood of transmission of other conjugative and mobilizable MGEs.

Coexistence of MGEs in the same cell can of course also result in the exchange of all sorts of genes among different elements, thus altering the potential availability of shuffled genes to new recipient hosts. In particular, novel properties affecting host range can appear by recombination among different types of MGEs. For instance, conjugative plasmids have been shown to appear by co-integration of ICEs and non-conjugative plasmids followed by deletions and other genomic rearrangements (Gasson *et al.*, 1992; Salyers *et al.*, 1995a,b). Similarly, ICEs can be turned into plasmids capable of sustaining their own replication by co-integration of plasmid-derived replication origins. For instance, *L. lactis* can harbour an ICE known as *sex factor*, whose transfer is generally limited to intraspecific strains. However, this ICE can be transmitted to a different Firmicutes genus if a plasmidic origin co-integrates in the circular intermediate that forms after excision from the host chromosome; the sex factor contains a group II intron which may in this way transfer between different genera (Belhocine *et al.*, 2007).

The shuffling of genes among MGEs is facilitated by several factors. First, most elements are organized in a highly modular fashion, as already mentioned above for plasmids and ICEs (Burrus *et al.*, 2002; Weaver *et al.*, 2009). Modularity and recombination also play a major role in phage evolution (Lucchini *et al.*, 1999) and recombination across MGEs of different types, via the exchange of modules or the co-integration of an element within another, is a common phenomenon. Plasmids, for instance, usually carry insertions of one or more transposons or other MGEs and the incorporation and expelling of such elements occur more frequently than do changes in the core plasmid backbone (Norberg *et al.*, 2011). For their part, ICEs and related mobilizable elements comprise functional modules that can be derived from phage, plasmids or transposons and it has been proposed that the combination of these modules generates, not discrete classes, but a continuum of mosaic MGEs (Osborn and Böltner, 2002). Further, MGEs can contain specialized systems for gene integration and expression, known as integrons. Integrons are genetic elements able to acquire and rearrange open reading frames (ORFs) embedded in gene cassette units and convert them to functional genes by ensuring their correct expression. They contain gene cassettes of heterogeneous origins probably collected successively in different genomics backgrounds. Integrons are rendered mobile by virtue of their association with transposons, often themselves carried on conjugative plasmids (Cambray *et al.*, 2010). In fact, transposon-associated integrons are one of the major conveyors of antibiotic multiresistance across Gram-negative and, to a lesser extent, Gram-positive bacteria (Partridge *et al.*, 2009). Non-mobile integrons are also present in the chromosomes of hundreds of bacterial species across diverse phyla and it is now thought, based on phylogenetic analysis, that the chromosomal integrons are the source of the integrons found in association with MGEs (Mazel *et al.*, 1998; Cambray *et al.*, 2010).

Overall relationship between phylogenetic distance and HGT

Although most gene transfers take place between closely related organisms (Lawrence and Hendrickson, 2003) with compatible gene contents, many studies have revealed that HGT events can and do occur between evolutionarily distant taxa. In a survey of curated protein domains in the Pfam database ($n = 7677$ families), Choi and Kim (2007) found that horizontal gene transfer decreased the higher the taxonomic range of the species considered regardless of the biological domain (Bacteria, Archaea or Eukarya). The extent of HGT varied from 1.1% (74 families) at the second, to 5.3% (365 families) at the third and 9.7% (667 families) at the fourth taxonomic ranges. In light of the wide breadth of taxonomic ranges actually covered in these three levels of the Pfam database (for instance, the third level comprised from phylum to family), it is not possible to derive estimates of HGT at recognizable taxonomic categories but, nevertheless, these authors concluded that HGT will have little impact on the construction of organism phylogenies, especially when phylogenomic methods or methods based on whole genome sequences are used.

Coscollá *et al.* (2011) studied what fraction of the genome of a single species results from strictly vertical inheritance and which derives from HGT events. The phylogenomic analysis of 1700 genes in four genome sequences of *Legionella pneumophila* revealed that as much as 42% of these genes were derived from non-γ-Proteobacteria, the class where *L. pneumophila* belongs. This large proportion of horizontally transferred genes is most likely facilitated by the intracellular, although transiently, lifestyle of *Legionella* in the cytoplasm of amoebas, which provide one of the environments most facilitating of lateral transfers (see above). In this same study, the authors analysed the relationship between number of HGT events and phylogenetic distance. Despite environmental facilitation, a clear negative correlation between these two variables was observed, thus indicating that horizontal transfers not only depend on opportunity but also on other limits some of which are clearly associated to evolutionary distance.

An interesting relationship between HGT from distant taxa, genome size, functional class and ecology has been suggested by Cordero and Hogeweg (2009). They analysed 333 complete bacterial genomes and determined the distribution of protein families in HGTs. They found that large genomes tended to have a larger contribution of evolutionarily distant genes, which were enriched in two functional categories, transcription regulation and defence mechanisms. The former category includes one-component systems, which incorporate DNA binding and substrate recognition domains, thus linking environmental stimuli and regulatory responses in a single protein. The most relevant component of HGT in the defence mechanisms corresponds to transporters involved in extrusion of toxic compounds and to R–M systems. In order to explain these patterns, Cordero and Hogeweg (2009) put forward a hypothesis in which complex environments would be selective for larger gene repertoires, in order to cope efficiently with the multiplicity of interactions in species-rich communities. The increase in gene repertoire seems to be mainly due to lateral transfer (Kuo and Ochman, 2009). The presence in these environments of phylogenetically diverse species along with the need to increase the gene repertoires of the genomes involved therein would lead to a circular process: more complex interactions in species-rich communities would increase the demand for new genes, which might be covered by incorporating transfers from distant species, resulting in more complex species and even more complex environments.

Conclusions

The flow of genes across taxa through HGT is enabled by the existence of complex networks of loopholes across transfer barriers, often provided by a small fraction of the individuals in an otherwise inaccessible population. HGT networks are fluid and highly evolvable, as is evidenced by the large intraspecific variability often observed in terms of the capacity of strains to donate or receive DNA by means of different mechanisms, and in the establishment of coevolutionary arms races among, particularly, bacteria and phage. One of the most important properties of HGT networks is their gene specificity, i.e. the fact that the genes of a potential donor genome will have an unequal likelihood of passing through or being incorporated in a given recipient. This gene specificity should have consequences for prokaryotic evolution at several different levels. From a gene-centric view, the long-term selective pressures that will operate on a gene will depend on the number and type of genomic backgrounds in which it might find itself and on the frequency of exchanges among them. At the other extreme, the existence of gene-specific transfer networks will create groups of organisms that will be partially related through gene sharing, but the relatedness among individuals will differ among genetic loci. Beyond the impact that they may have on prokaryotic phylogeny and species concepts, the specific ways in which HGT networks will affect genic, organismal and phyletic evolution need to be quantitatively explored, as they could result in unforeseen evolutionary dynamics in the prokaryotic world.

Acknowledgement

We would like to thank Yvonne Vallès for her help with the references cited in this chapter.

References

Abedon, S.T. (2009). Kinetics of phage-mediated biocontrol of bacteria. Foodborne Pathog. Dis. *6*, 807–815.

Amitsur, M., Levitz, R., and Kaufmann, G. (1987). Bacteriophage T4 anticodon nuclease, polynucleotide kinase and RNA ligase reprocess the host lysine tRNA. EMBO J. *6*, 2499–2503.

Andersson, A.F., and Banfield, J.F. (2008). Virus population dynamics and acquired virus resistance in natural microbial communities. Science *320*, 1047–1050.

Aras, R.A., Takata, T., Ando, T., van der Ende, A., and Blaser, M.J. (2001). Regulation of the HpyII restriction-modification system of *Helicobacter pylori* by gene deletion and horizontal reconstitution. Mol. Microbiol. *42*, 369–382.

Atanasiu, C., Byron, O., McMiken, H., Sturrock, S.S., and Dryden, D.T. (2001). Characterisation of the structure of ocr, the gene 0.3 protein of bacteriophage T7. Nucleic Acids Res. *29*, 3059–3068.

Audic, S., Robert, C., Campagna, B., Parinello, H., Claverie, J.M., Raoult, D., and Drancourt, M. (2007). Genome analysis of *Minibacterium massiliensis* highlights the convergent evolution of water-living bacteria. PLoS Genet. *3*, e138.

Averhoff, B., and Friedrich, A. (2003). Type IV pili-related natural transformation systems: DNA transport in mesophilic and thermophilic bacteria. Arch. Microbiol. *180*, 385–393.

Bandyopadhyay, P.K., Studier, F.W., Hamilton, D.L., and Yuan, R. (1985). Inhibition of the type I restriction-modification enzymes *EcoB* and *EcoK* by the gene 0.3 protein of bacteriophage T7. J. Mol. Biol. *182*, 567–578.

Bapteste, E., and Boucher, Y. (2009). Epistemological impacts of horizontal gene transfer on classification in microbiology. Methods Mol. Biol. *532*, 55–72.

Barrangou, R., Fremaux, C., Deveau, H., Richards, M., Boyaval, P., Moineau, S., Romero, D.A., and Horvath, P. (2007). CRISPR provides acquired resistance against viruses in prokaryotes. Science *315*, 1709–1712.

Bates, S., Roscoe, R.A., Althorpe, N.J., Brammar, W.J., and Wilkins, B.M. (1999). Expression of leading region genes on lncl1 plasmid Collb-P9: genetic evidence for single-stranded DNA transcription. Microbiology *145*, 2655-2662.

Bayliss, C.D., Callaghan, M.J., and Moxon, E.R. (2006). High allelic diversity in the methyltransferase gene of a phase variable type III restriction-modification system has implications for the fitness of *Haemophilus influenzae*. Nucleic Acids Res. *34*, 4046–4059.

Beaber, J.W., Burrus, V., Hochhut, B., and Waldor, M.K. (2002). Comparison of SXT and R391, two conjugative integrating elements: definition of a genetic backbone for the mobilization of resistance determinants. Cell. Mol. Life Sci. *59*, 2065–2070.

Beiko, R.G., Harlow, T.J., and Ragan, M.A. (2005). Highways of gene sharing in prokaryotes. Proc. Natl. Acad. Sci. U.S.A. *102*, 14332–14337.

Belhocine, K., Mandilaras, V., Yeung, B., and Cousineau, B. (2007). Conjugative transfer of the *Lactococcus lactis* sex factor and pRS01 plasmid to *Enterococcus faecalis*. FEMS Microbiol. Lett. *269*, 289–294.

Berndt, C., Meier, P., and Wackernagel, W. (2003). DNA restriction is a barrier to natural transformation in *Pseudomonas stutzeri* JM300. Microbiology *149*, 895–901.

Beumer, A., and Robinson, J.B. (2005). A broad-host-range, generalized transducing phage (SN-T) acquires 16S rRNA genes from different genera of bacteria. Appl. Environ. Microbiol. *71*, 8301–8304.

Bianchi, M.E., and Radding, C.M. (1983). Insertions, deletions and mismatches in heteroduplex DNA made by recA protein. Cell *35*, 511–520.

Biswas, G.D., and Sparling, P.F. (1981). Entry of double-stranded deoxyribonucleic acid during transformation of *Neisseria gonorrhoeae*. J. Bacteriol. *145*, 638–640.

Bordenstein, S.R., and Reznikoff, W.S. (2005). Mobile DNA in obligate intracellular bacteria. Nat. Rev. Microbiol. *3*, 688–699.

Boucher, Y., Huber, H., L'Haridon, S., Stetter, K.O., and Doolittle, W.F. (2001). Bacterial origin for the isoprenoid biosynthesis enzyme HMG-CoA reductase of the archaeal orders Thermoplasmatales and Archaeoglobales. Mol. Biol. Evol. *18*, 1378–1388.

Breitbart, M., and Rohwer, F. (2005). Here a virus, there a virus, everywhere the same virus? Trends Microbiol. *13*, 278–284.

Brigulla, M., and Wackernagel, W. (2010). Molecular aspects of gene transfer and foreign DNA acquisition in prokaryotes with regard to safety issues. Appl. Microbiol. Biotechnol. *86*, 1027–1041.

Budroni, S., Siena, E., Dunning Hotopp, J.C., Seib, K.L., Serruto, D., Nofroni, C., Comanducci, M., Riley, D.R., Daugherty, S.C., Angiuoli, S.V., *et al.* (2011). *Neisseria meningitidis* is structured in clades associated with restriction modification systems that modulate homologous recombination. Proc. Natl. Acad. Sci. U.S.A. *108*, 4494–4499.

Bujnicki, J.M. (2001). Understanding the evolution of restriction-modification systems: clues from sequence and structure comparisons. Acta Biochim. Pol. *48*, 935–967.

Burrus, V., Pavlovic, G., Decaris, B., and Guedon, G. (2002). Conjugative transposons: the tip of the iceberg. Mol. Microbiol. *46*, 601–610.

Cambray, G., Guerout, A.M., and Mazel, D. (2010). Integrons. Annu. Rev. Genet. *44*, 141–166.

Caspi, R., Helinski, D.R., Pacek, M., and Konieczny, I. (2000). Interactions of DnaA proteins from distantly related bacteria with the replication origin of the broad host range plasmid RK2. J. Biol. Chem. *275*, 18454–18461.

Charpentier, X., Kay, E., Schneider, D., and Shuman, H.A. (2011). Antibiotics and UV radiation induce competence for natural transformation in *Legionella pneumophila*. J. Bacteriol. *193*, 1114–1121.

Chen, I., and Dubnau, D. (2003). DNA transport during transformation. Front. Biosci. *8*, s544–556.

Chen, J.M., Chuzhanova, N., Stenson, P.D., Ferec, C., and Cooper, D.N. (2005). Intrachromosomal serial replication slippage in trans gives rise to diverse genomic rearrangements involving inversions. Hum. Mutat. *26*, 362–373.

Choi, I.G., and Kim, S.H. (2007). Global extent of horizontal gene transfer. Proc. Natl. Acad. Sci. U.S.A. *104*, 4489–4494.

Clewell, D.B. (1993). Bacterial sex pheromone-induced plasmid transfer. Cell 73, 9–12.

Clewell, D.B., Jaworski, D.D., Flannagan, S.E., Zitzow, L.A., and Su, Y.A. (1995). The conjugative transposon Tn916 of *Enterococcus faecalis*: structural analysis and some key factors involved in movement. Dev. Biol. Stand. *85*, 11–17.

Cordero, O.X., and Hogeweg, P. (2009). The impact of long-distance horizontal gene transfer on prokaryotic genome size. Proc. Natl. Acad. Sci. U.S.A. *106*, 21748–21753.

Corvaglia, A.R., Francois, P., Hernandez, D., Perron, K., Linder, P., and Schrenzel, J. (2010). A type III-like restriction endonuclease functions as a major barrier to horizontal gene transfer in clinical *Staphylococcus aureus* strains. Proc. Natl. Acad. Sci. U.S.A. *107*, 11954–11958.

Coscollá, M., Comas, I., and González-Candelas, F. (2011). Quantifying nonvertical inheritance in the evolution of *Legionella pneumophila*. Mol. Biol. Evol. *28*, 985–1001.

De Gelder, L., Vandecasteele, F.P., Brown, C.J., Forney, L.J., and Top, E.M. (2005). Plasmid donor affects host range of promiscuous IncP-1beta plasmid pB10 in an activated-sludge microbial community. Appl. Environ. Microbiol. *71*, 5309–5317.

De Gelder, L., Ponciano, J.M., Joyce, P., and Top, E.M. (2007). Stability of a promiscuous plasmid in different hosts: no guarantee for a long–term relationship. Microbiology *153*, 452–463.

De Gelder, L., Williams, J.J., Ponciano, J.M., Sota, M., and Top, E.M. (2008). Adaptive plasmid evolution results in host-range expansion of a broad-host-range plasmid. Genetics *178*, 2179–2190.

Deane, S.M., and Rawlings, D.E. (2004). Plasmid evolution and interaction between the plasmid addiction stability systems of two related broad-host-range IncQ-like plasmids. J. Bacteriol. *186*, 2123–2133.

DeMars, R., Weinfurter, J., Guex, E., Lin, J., and Potucek, Y. (2007). Lateral gene transfer *in vitro* in the intracellular pathogen *Chlamydia trachomatis*. J. Bacteriol. *189*, 991–1003.

Diaz, R., and Staudenbauer, W.L. (1982). Replication of the broad host range plasmid RSF1010 in cell-free extracts of *Escherichia coli* and *Pseudomonas aeruginosa*. Nucleic Acids Res. *10*, 4687–4702.

Doolittle, W.F., and Zhaxybayeva, O. (2009). On the origin of prokaryotic species. Genome Res. *19*, 744–756.

Doran, K.S., Helinski, D.R., and Konieczny, I. (1999). Host-dependent requirement for specific DnaA boxes for plasmid RK2 replication. Mol. Microbiol. *33*, 490–498.

Dreiseikelmann, B. (1994). Translocation of DNA across bacterial membranes. Microbiol. Rev. *58*, 293–316.

Droge, M., Puhler, A., and Selbitschka, W. (1998). Horizontal gene transfer as a biosafety issue: a natural phenomenon of public concern. J. Biotechnol. *64*, 75–90.

Dunning Hotopp, J. (2011). Horizontal gene transfer between bacteria and animals. Trends Genet. *27*, 157–163.

Dybvig, K., Sitaraman, R., and French, C.T. (1998). A family of phase-variable restriction enzymes with differing specificities generated by high-frequency gene rearrangements. Proc. Natl. Acad. Sci. U.S.A. *95*, 13923–13928.

Fricke, W.F., Mammel, M.K., McDermott, P.F., Tartera, C., White, D.G., Leclerc, J.E., Ravel, J., and Cebula, T.A. (2011). Comparative genomics of 28 *Salmonella enterica* isolates: evidence for CRISPR-mediated adaptive sublineage evolution. J. Bacteriol. *193*, 3556–3568.

García de Viedma, D., Giraldo, R., Rivas, G., Fernández-Tresguerres, E., and Díaz-Orejas, R. (1996). A leucine zipper motif determines different functions in a DNA replication protein. EMBO J. *15*, 925–934.

Garcillán-Bárcia, M.P., and de la Cruz, F. (2008). Why is entry exclusion an essential feature of conjugative plasmids? Plasmid *60*, 1–18.

Garcillán-Barcia, M.P., Jurado, P., González-Perez, B., Moncalián, G., Fernández, L.A., and de la Cruz, F. (2007). Conjugative transfer can be inhibited by blocking relaxase activity within recipient cells with intrabodies. Mol. Microbiol. *63*, 404–416.

Gasson, M.J., Swindell, S., Maeda, S., and Dodd, H.M. (1992). Molecular rearrangement of lactose plasmid DNA associated with high-frequency transfer and cell aggregation in *Lactococcus lactis* 712. Mol. Microbiol. *6*, 3213–3223.

Ghigo, J.M. (2001). Natural conjugative plasmids induce bacterial biofilm development. Nature *412*, 442–445.

Gillespie, J.J., Williams, K., Shukla, M., Snyder, E.E., Nordberg, E.K., Ceraul, S.M., Dharmanolla, C., Rainey, D., Soneja, J., Shallom, J.M., *et al.* (2008). *Rickettsia* phylogenomics: unwinding the intricacies of obligate intracellular life. PLoS One *3*, e2018.

Glansdorff, N., Xu, Y., and Labedan, B. (2008). The last universal common ancestor: emergence, constitution and genetic legacy of an elusive forerunner. Biol. Direct *3*, 29.

Godon, J.J., Jury, K., Shearman, C.A., and Gasson, M.J. (1994). The *Lactococcus lactis* sex-factor aggregation gene *cluA*. Mol. Microbiol. *12*, 655–663.

Goldberg, S., and Murphy, J.R. (1983). Molecular epidemiological studies of United States Gulf Coast *Vibrio cholerae* strains: integration site of mutator vibriophage VcA-3. Infect. Immun. *42*, 224–230.

Gormley, E.P., and Davies, J. (1991). Transfer of plasmid RSF1010 by conjugation from *Escherichia coli* to *Streptomyces lividans* and *Mycobacterium smegmatis*. J. Bacteriol. *173*, 6705–6708.

Grant, A.J., Bird, P.I., and Pittard, J. (1980). Naturally occurring plasmids exhibiting incompatibility with members of incompatibility groups I and P. J. Bacteriol. *144*, 758–765.

Guiney, D.G. (1982). Host range of conjugation and replication functions of the *Escherichia coli* sex plasmid *Flac*. Comparison with the broad host-range plasmid RK2. J. Mol. Biol. *162*, 699–703.

Haaber, J., Moineau, S., and Hammer, K. (2009). Activation and transfer of the chromosomal phage resistance mechanism AbiV in *Lactococcus lactis*. Appl. Environ. Microbiol. 75 3358–3361.

Halary, S., Leigh, J.W., Cheaib, B., Lopez, P., and Bapteste, E. (2010). Network analyses structure genetic diversity in independent genetic worlds. Proc. Natl. Acad. Sci. U.S.A. *107*, 127–132.

Hanfling, P., Shashkov, A.S., Jann, B., and Jann, K. (1996). Analysis of the enzymatic cleavage (beta elimination) of the capsular K5 polysaccharide of *Escherichia coli* by the K5-specific coliphage: reexamination. J. Bacteriol. *178*, 4747–4750.

Harms, K., and Wackernagel, W. (2008). The RecBCD and SbcCD DNases suppress homology-facilitated illegitimate recombination during natural transformation of *Acinetobacter baylyi*. Microbiology *154*, 2437–2445.

Havarstein, L.S., Coomaraswamy, G., and Morrison, D.A. (1995). An unmodified heptadecapeptide pheromone induces competence for genetic transformation in *Streptococcus pneumoniae*. Proc. Natl. Acad. Sci. U.S.A. *92*, 11140–11144.

Hecht, M.M., Nitz, N., Araujo, P.F., Sousa, A.O., Rosa Ade, C., Gomes, D.A., Leonardecz, E., and Teixeira, A.R. (2010). Inheritance of DNA transferred from American trypanosomes to human hosts. PLoS One *5*, e9181.

Heuer, H., Fox, R.E., and Top, E.M. (2007). Frequent conjugative transfer accelerates adaptation of a broad-host-range plasmid to an unfavorable *Pseudomonas putida* host. FEMS Microbiol. Ecol. *59*, 738–748.

Hilario, E., and Gogarten, J.P. (1993). Horizontal transfer of ATPase genes – the tree of life becomes a net of life. Biosystems *31*, 111–119.

Holmfeldt, K., Middelboe, M., Nybroe, O., and Riemann, L. (2007). Large variabilities in host strain susceptibility and phage host range govern interactions between lytic marine phages and their *Flavobacterium* hosts. Appl. Environ. Microbiol. *73*, 6730–6739.

Hoskisson, P.A., and Smith, M.C. (2007). Hypervariation and phase variation in the bacteriophage 'resistome'. Curr. Opin. Microbiol. *10*, 396–400.

Hulter, N., and Wackernagel, W. (2008). Frequent integration of short homologous DNA tracks during *Acinetobacter baylyi* transformation and influence of transcription and RecJ and SbcCD DNases. Microbiology *154*, 3676–3685.

Hussain, H.A., Roberts, A.P., and Mullany, P. (2005). Generation of an erythromycin-sensitive derivative of *Clostridium difficile* strain 630 (630Δ*erm*) and demonstration that the conjugative transposon Tn*916*ΔE enters the genome of this strain at multiple sites. J. Med. Microbiol. *54*, 137–141.

Hyman, P., and Abedon, S.T. (2010). Bacteriophage host range and bacterial resistance. Adv. Appl. Microbiol. *70*, 217–248.

Ikemura, T. (1985). Codon usage and tRNA content in unicellular and multicellular organisms. Mol. Biol. Evol. *2*, 13–34.

Itaya, M. (1999). Genetic transfer of large DNA inserts to designated loci of the *Bacillus subtilis* 168 genome. J. Bacteriol. *181*, 1045–1048.

Jain, R., Rivera, M.C., Moore, J.E., and Lake, J.A. (2003). Horizontal gene transfer accelerates genome innovation and evolution. Mol. Biol. Evol. *20*, 1598–1602.

Jensen, E.C., Schrader, H.S., Rieland, B., Thompson, T.L., Lee, K.W., Nickerson, K.W., and Kokjohn, T.A. (1998). Prevalence of broad-host-range lytic bacteriophages of *Sphaerotilus natans, Escherichia coli,* and *Pseudomonas aeruginosa*. Appl. Environ. Microbiol. *64*, 575–580.

Johnsborg, O., Eldholm, V., Bjornstad, M.L., and Havarstein, L.S. (2008). A predatory mechanism dramatically increases the efficiency of lateral gene transfer in *Streptococcus pneumoniae* and related commensal species. Mol. Microbiol. *69*, 245–253.

Jyssum, K., Jyssum, S., and Gundersen, W.B. (1971). Sorption of DNA and RNA during transformation of *Neisseria meningitidis*. Acta Pathol. Microbiol. Scand. B Microbiol. Immunol. *79*, 563–571.

Kanaya, S., Yamada, Y., Kudo, Y., and Ikemura, T. (1999). Studies of codon usage and tRNA genes of 18 unicellular organisms and quantification of *Bacillus subtilis* tRNAs: gene expression level and species-specific diversity of codon usage based on multivariate analysis. Gene *238*, 143–155.

Keeling, P.J., and Palmer, J.D. (2008). Horizontal gene transfer in eukaryotic evolution. Nat. Rev. Genet. *9*, 605–618.

Kenzaka, T., Tani, K., and Nasu, M. (2010). High-frequency phage-mediated gene transfer in freshwater environments determined at single-cell level. ISME J. *4*, 648–659.

Kleter, G.A., Peijnenburg, A.A., and Aarts, H.J. (2005). Health considerations regarding horizontal transfer of microbial transgenes present in genetically modified crops. J. Biomed. Biotechnol. *2005*, 326–352.

Koonin, E.V., and Wolf, Y.I. (2009). The fundamental units, processes and patterns of evolution, and the tree of life conundrum. Biol. Direct *4*, 33.

Kramer, M.G., Khan, S.A., and Espinosa, M. (1997). Plasmid rolling circle replication: identification of the RNA polymerase-directed primer RNA and requirement for DNA polymerase I for lagging strand synthesis. EMBO J. *16*, 5784–5795.

Kreth, J., Merritt, J., Shi, W., and Qi, F. (2005). Competition and coexistence between *Streptococcus mutans* and *Streptococcus sanguinis* in the dental biofilm. J. Bacteriol. *187*, 7193–7203.

Krzywinska, E., Krzywinski, J., and Schorey, J.S. (2004). Naturally occurring horizontal gene transfer and homologous recombination in *Mycobacterium*. Microbiology *150*, 1707–1712.

Kuo, C.H., and Ochman, H. (2009). The fate of new bacterial genes. FEMS Microbiol. Rev. *33*, 38–43.

La Scola, B., Desnues, C., Pagnier, I., Robert, C., Barrassi, L., Fournous, G., Merchat, M., Suzan-Monti, M., Forterre, P., Koonin, E., *et al.* (2008). The virophage as a unique parasite of the giant mimivirus. Nature *455*, 100–104.

Labrie, S.J., and Moineau, S. (2007). Abortive infection mechanisms and prophage sequences significantly influence the genetic makeup of emerging lytic lactococcal phages. J. Bacteriol. *189*, 1482–1487.

Lacks, S.A., and Springhorn, S.S. (1984). Transfer of recombinant plasmids containing the gene for DpnII DNA methylase into strains of *Streptococcus pneumoniae* that produce DpnI or DpnII restriction endonucleases. J. Bacteriol. *158*, 905–909.

Langley, R., Kenna, D.T., Vandamme, P., Ure, R., and Govan, J.R. (2003). Lysogeny and bacteriophage host range within the *Burkholderia cepacia* complex. J. Med. Microbiol. *52*, 483–490.

Lawrence, J.G., and Hendrickson, H. (2003). Lateral gene transfer: when will adolescence end? Mol. Microbiol. *50*, 739–749.

Lee, C.A., and Grossman, A.D. (2007). Identification of the origin of transfer (oriT) and DNA relaxase required for conjugation of the integrative and conjugative element ICEBs1 of *Bacillus subtilis*. J. Bacteriol. *189*, 7254–7261.

Lee, C.H., Liu, J.W., Li, C.C., Chien, C.C., Tang, Y.F., and Su, L.H. (2011). Spread of ISCR1 elements containing *bla*DHA-1 and multiple antimicrobial resistance genes leading to increase of flomoxef resistance in extended-spectrum-beta-lactamase-producing *Klebsiella pneumoniae*. Antimicrob. Agents Chemother. *55*, 4058–4063.

Leiman, P.G., Battisti, A.J., Bowman, V.D., Stummeyer, K., Muhlenhoff, M., Gerardy-Schahn, R., Scholl, D., and Molineux, I.J. (2007). The structures of bacteriophages K1E and K1-5 explain processive degradation of polysaccharide capsules and evolution of new host specificities. J. Mol. Biol. *371*, 836–849.

Levine, S.M., Lin, E.A., Emara, W., Kang, J., DiBenedetto, M., Ando, T., Falush, D., and Blaser, M.J. (2007). Plastic cells and populations: DNA substrate characteristics in *Helicobacter pylori* transformation define a flexible but conservative system for genomic variation. FASEB J. *21*, 3458–3467.

van der Ley, P., de Graaff, P., and Tommassen, J. (1986). Shielding of *Escherichia coli* outer membrane proteins as receptors for bacteriophages and colicins by O-antigenic chains of lipopolysaccharide. J. Bacteriol. *168*, 449–451.

Liu, F., Kariyawasam, S., Jayarao, B.M., Barrangou, R., Gerner-Smidt, P., Ribot, E.M., Knabel, S.J., and Dudley, E.G. (2011). Subtyping *Salmonella enterica* serovar enteritidis isolates from different sources by using sequence typing based on virulence genes and clustered regularly interspaced short palindromic repeats (CRISPRs). Appl. Environ. Microbiol. *77*, 4520–4526.

Liu, M., Deora, R., Doulatov, S.R., Gingery, M., Eiserling, F.A., Preston, A., Maskell, D.J., Simons, R.W., Cotter, P.A., Parkhill, J., *et al.* (2002). Reverse transcriptase-mediated tropism switching in *Bordetella* bacteriophage. Science *295*, 2091–2094.

Llosa, M., Gomis-Ruth, F.X., Coll, M., and de la Cruz Fd, F. (2002). Bacterial conjugation: a two-step mechanism for DNA transport. Mol. Microbiol. *45*, 1–8.

Lorenz, M.G., and Wackernagel, W. (1994). Bacterial gene transfer by natural genetic transformation in the environment. Microbiol. Rev. *58*, 563–602.

Lucchini, S., Desiere, F., and Brussow, H. (1999). The genetic relationship between virulent and temperate *Streptococcus thermophilus* bacteriophages: whole genome comparison of *cos*-site phages Sfi19 and Sfi21. Virology *260*, 232–243.

Lucchini, S., Rowley, G., Goldberg, M.D., Hurd, D., Harrison, M., and Hinton, J.C. (2006). H-NS mediates the silencing of laterally acquired genes in bacteria. PLoS Pathog. *2*, e81.

McDaniel, L.D., Young, E., Delaney, J., Ruhnau, F., Ritchie, K.B., and Paul, J.H. (2010). High frequency of horizontal gene transfer in the oceans. Science *330*, 50.

McMahon, S.A., Roberts, G.A., Johnson, K.A., Cooper, L.P., Liu, H., White, J.H., Carter, L.G., Sanghvi, B., Oke, M., Walkinshaw, M.D., *et al.* (2009). Extensive DNA mimicry by the ArdA anti-restriction protein and its role in the spread of antibiotic resistance. Nucleic Acids Res. *37*, 4887–4897.

Majewski, J., Zawadzki, P., Pickerill, P., Cohan, F.M., and Dowson, C.G. (2000). Barriers to genetic exchange between bacterial species: *Streptococcus pneumoniae* transformation. J. Bacteriol. *182*, 1016–1023.

Makarova, K.S., Haft, D.H., Barrangou, R., Brouns, S.J., Charpentier, E., Horvath, P., Moineau, S., Mojica, F.J., Wolf, Y.I., Yakunin, A.F., *et al.* (2011). Evolution and classification of the CRISPR-Cas systems. Nat. Rev. Microbiol. *9*, 467–477.

Marraffini, L.A., and Sontheimer, E.J. (2008). CRISPR interference limits horizontal gene transfer in staphylococci by targeting DNA. Science *322*, 1843–1845.

Mathee, K., Narasimhan, G., Valdes, C., Qiu, X., Matewish, J.M., Koehrsen, M., Rokas, A., Yandava, C.N., Engels, R., Zeng, E., *et al.* (2008). Dynamics of *Pseudomonas aeruginosa* genome evolution. Proc. Natl. Acad. Sci. U.S.A. *105*, 3100–3105.

Matic, I., Rayssiguier, C., and Radman, M. (1995). Interspecies gene exchange in bacteria: the role of SOS and mismatch repair systems in evolution of species. Cell *80*, 507–515.

Matic, I., Taddei, F., and Radman, M. (1996). Genetic barriers among bacteria. Trends Microbiol. *4*, 69–72.

Mazel, D., Dychinco, B., Webb, V.A., and Davies, J. (1998). A distinctive class of integron in the *Vibrio cholerae* genome. Science *280*, 605–608.

Medhekar, B., and Miller, J.F. (2007). Diversity-generating retroelements. Curr. Opin. Microbiol. *10*, 388–395.

Mehr, I.J., and Seifert, H.S. (1998). Differential roles of homologous recombination pathways in *Neisseria gonorrhoeae* pilin antigenic variation, DNA transformation and DNA repair. Mol. Microbiol. *30*, 697–710.

Meier, P., and Wackernagel, W. (2005). Impact of *mutS* inactivation on foreign DNA acquisition by natural transformation in *Pseudomonas stutzeri*. J. Bacteriol. *187*, 143–154.

Merryweather, A., Barth, P.T., and Wilkins, B.M. (1986). Role and specificity of plasmid RP4-encoded DNA primase in bacterial conjugation. J. Bacteriol. *167*, 12–17.

Meyer, R. (2009). Replication and conjugative mobilization of broad host-range IncQ plasmids. Plasmid *62*, 57–70.

Milkman, R., Raleigh, E.A., McKane, M., Cryderman, D., Bilodeau, P., and McWeeny, K. (1999). Molecular evolution of the *Escherichia coli* chromosome. V. Recombination patterns among strains of diverse origin. Genetics *153*, 539–554.

Mojica, F.J., and Díez-Villaseñor, C. (2010). The on–off switch of CRISPR immunity against phages in *Escherichia coli*. Mol. Microbiol. *77*, 1341–1345.

Mojica, F.J., Díez-Villasenñor, C., García-Martínez, J., and Soria, E. (2005). Intervening sequences of regularly spaced prokaryotic repeats derive from foreign genetic elements. J. Mol. Evol. *60*, 174–182.

Montag, D., Hashemolhosseini, S., and Henning, U. (1990). Receptor-recognizing proteins of T-even type bacteriophages. The receptor-recognizing area of proteins 37 of phages T4 TuIa and TuIb. J. Mol. Biol. *216*, 327–334.

Morris, P., Marinelli, L.J., Jacobs-Sera, D., Hendrix, R.W., and Hatfull, G.F. (2008). Genomic characterization of mycobacteriophage Giles: evidence for phage acquisition of host DNA by illegitimate recombination. J. Bacteriol. *190*, 2172–2182.

Murray, N.E. (2000). Type I restriction systems: sophisticated molecular machines (a legacy of Bertani and Weigle). Microbiol. Mol. Biol. Rev. *64*, 412–434.

Murray, N.E. (2002). 2001 Fred Griffith review lecture. Immigration control of DNA in bacteria: self versus non-self. Microbiology *148*, 3–20.

Musovic, S., Oregaard, G., Kroer, N., and Sorensen, S.J. (2006). Cultivation-independent examination of horizontal transfer and host range of an IncP-1 plasmid among Gram-positive and Gram-negative bacteria indigenous to the barley rhizosphere. Appl. Environ. Microbiol. *72*, 6687–6692.

Navarre, W.W., McClelland, M., Libby, S.J., and Fang, F.C. (2007). Silencing of xenogeneic DNA by H-NS-facilitation of lateral gene transfer in bacteria by a defense system that recognizes foreign DNA. Genes Dev. *21*, 1456–1471.

Navarre, W.W., Porwollik, S., Wang, Y., McClelland, M., Rosen, H., Libby, S.J., and Fang, F.C. (2006). Selective silencing of foreign DNA with low GC content by the H-NS protein in *Salmonella*. Science *313*, 236–238.

Nelson, K.E., Clayton, R.A., Gill, S.R., Gwinn, M.L., Dodson, R.J., Haft, D.H., Hickey, E.K., Peterson, J.D., Nelson, W.C., Ketchum, K.A., *et al.* (1999). Evidence for lateral gene transfer between Archaea and bacteria from genome sequence of *Thermotoga maritima*. Nature *399*, 323–329.

Nesvera, J., Hochmannova, J., Patek, M., Sroglova, A., and Becvarova, V. (1994). Transfer of the broad-host-range IncQ plasmid RSF1010 and other plasmid vectors to the Gram-positive methylotroph *Brevibacterium methylicum* by electrotransformation. Appl. Microbiol. Biotechnol. *40*, 864–866.

Nielsen, K.M., Johnsen, P.J., Bensasson, D., and Daffonchio, D. (2007). Release and persistence of extracellular DNA in the environment. Environ. Biosafety Res. *6*, 37–53.

Nishigaki, K., Kaneko, Y., Wakuda, H., Husimi, Y., and Tanaka, T. (1985). Type II restriction endonucleases cleave single-stranded DNAs in general. Nucleic Acids Res. *13*, 5747–5760.

Norberg, P., Bergstrom, M., Jethava, V., Dubhashi, D., and Hermansson, M. (2011). The IncP-1 plasmid backbone adapts to different host bacterial species and evolves through homologous recombination. Nat. Commun. *2*, 268.

Norman, A., Hansen, L.H., She, Q., and Sorensen, S.J. (2008). Nucleotide sequence of pOLA52: a conjugative IncX1 plasmid from *Escherichia coli* which enables biofilm formation and multidrug efflux. Plasmid *60*, 59–74.

Norman, A., Hansen, L.H., and Sorensen, S.J. (2009). Conjugative plasmids: vessels of the communal gene pool. Philos. Trans. R Soc. Lond. B Biol. Sci. *364*, 2275–2289.

Nzula, S., Vandamme, P., and Govan, J.R. (2000). Sensitivity of the *Burkholderia cepacia* complex and *Pseudomonas aeruginosa* to transducing bacteriophages. FEMS Immunol. Med. Microbiol. *28*, 307–312.

O'Driscoll, J., Glynn, F., Fitzgerald, G.F., and van Sinderen, D. (2006). Sequence analysis of the lactococcal plasmid pNP40: a mobile replicon for coping with environmental hazards. J. Bacteriol. *188*, 6629–6639.

Osborn, A.M., and Böltner, D. (2002). When phage, plasmids, and transposons collide: genomic islands, and conjugative- and mobilizable-transposons as a mosaic continuum. Plasmid *48*, 202–212.

Oshima, T., Ishikawa, S., Kurokawa, K., Aiba, H., and Ogasawara, N. (2006). *Escherichia coli* histone-like protein H-NS preferentially binds to horizontally acquired DNA in association with RNA polymerase. DNA Res. *13*, 141–153.

Pál, C., Papp, B., and Lercher, M.J. (2005). Horizontal gene transfer depends on gene content of the host. Bioinformatics *21(Suppl. 2)*, ii222–223.

Pallen, M.J., and Wren, B.W. (2007). Bacterial pathogenomics. Nature *449*, 835–842.

Partridge, S.R., Tsafnat, G., Coiera, E., and Iredell, J.R. (2009). Gene cassettes and cassette arrays in mobile resistance integrons. FEMS Microbiol. Rev. *33*, 757–784.

Pelkonen, S., Aalto, J., and Finne, J. (1992). Differential activities of bacteriophage depolymerase on bacterial polysaccharide: binding is essential but degradation is inhibitory in phage infection of K1-defective *Escherichia coli*. J. Bacteriol. *174*, 7757–7761.

Pérez-Mendoza, D., and de la Cruz, F. (2009). *Escherichia coli* genes affecting recipient ability in plasmid conjugation: are there any? BMC Genomics *10*, 71.

Perry, L.L., SanMiguel, P., Minocha, U., Terekhov, A.I., Shroyer, M.L., Farris, L.A., Bright, N., Reuhs, B.L., and Applegate, B.M. (2009). Sequence analysis of *Escherichia coli* O157:H7 bacteriophage PhiV10 and identification of a phage-encoded immunity protein that modifies the O157 antigen. FEMS Microbiol. Lett. *292*, 182–186.

Pinkney, M., Diaz, R., Lanka, E., and Thomas, C.M. (1988). Replication of mini RK2 plasmid in extracts of *Escherichia coli* requires plasmid-encoded protein TrfA and host-encoded proteins DnaA, B, G DNA gyrase and DNA polymerase III. J. Mol. Biol. *203*, 927–938.

Pride, D.T., Salzman, J., Haynes, M., Rohwer, F., Davis-Long, C., White, R.A., 3rd, Loomer, P., Armitage, G.C., and Relman, D.A. (2012). Evidence of a robust resident bacteriophage population revealed through analysis of the human salivary virome. ISME J. *6*, 915-926.

Puigbo, P., Wolf, Y.I., and Koonin, E.V. (2009). Search for a 'Tree of Life' in the thicket of the phylogenetic forest. J. Biol. *8*, 59.

Pul, U., Wurm, R., Arslan, Z., Geissen, R., Hofmann, N., and Wagner, R. (2010). Identification and characterization of *E. coli* CRISPR-cas promoters and their silencing by H-NS. Mol. Microbiol. *75*, 1495–1512.

Ragan, M.A., and Beiko, R.G. (2009). Lateral genetic transfer: open issues. Philos. Trans. R. Soc. Lond. B Biol. Sci. *364*, 2241–2251.

Rawlings, D.E., and Tietze, E. (2001). Comparative biology of IncQ and IncQ-like plasmids. Microbiol. Mol. Biol. Rev. *65*, 481–496.

Rayssiguier, C., Thaler, D.S., and Radman, M. (1989). The barrier to recombination between *Escherichia coli* and *Salmonella typhimurium* is disrupted in mismatch-repair mutants. Nature *342*, 396–401.

Renesto, P., Ogata, H., Audic, S., Claverie, J.M., and Raoult, D. (2005). Some lessons from *Rickettsia* genomics. FEMS Microbiol. Rev. *29*, 99–117.

Riley, M.A., and Lizotte-Waniewski, M. (2009). Population genomics and the bacterial species concept. Methods Mol. Biol. *532*, 367–377.

Roberts, A.P., Chandler, M., Courvalin, P., Guedon, G., Mullany, P., Pembroke, T., Rood, J.I., Smith, C.J., Summers, A.O., Tsuda, M., *et al.* (2008). Revised nomenclature for transposable genetic elements. Plasmid *60*, 167–173.

Rocha, E.P. (2004). Codon usage bias from tRNA's point of view: redundancy, specialization, and efficient decoding for translation optimization. Genome Res. *14*, 2279–2286.

Rocha, E.P., and Danchin, A. (2002). Base composition bias might result from competition for metabolic resources. Trends Genet. *18*, 291–294.

Salyers, A.A., Shoemaker, N.B., and Li, L.Y. (1995a). In the driver's seat: the *Bacteroides* conjugative transposons and the elements they mobilize. J. Bacteriol. *177*, 5727–5731.

Salyers, A.A., Shoemaker, N.B., Stevens, A.M., and Li, L.Y. (1995b). Conjugative transposons: an unusual and diverse set of integrated gene transfer elements. Microbiol. Rev. *59*, 579–590.

Saunders, C.W., and Guild, W.R. (1981). Monomer plasmid DNA transforms *Streptococcus pneumoniae*. Mol. Gen. Genet. *181*, 57–62.

Saunders, N.J., Hood, D.W., and Moxon, E.R. (1999). Bacterial evolution: bacteria play pass the gene. Curr. Biol. *9*, R180–183.

Scherzinger, E., Haring, V., Lurz, R., and Otto, S. (1991). Plasmid RSF1010 DNA replication *in vitro* promoted by purified RSF1010 RepA, RepB and RepC proteins. Nucleic Acids Res. *19*, 1203–1211.

Scholl, D., and Merril, C. (2005). The genome of bacteriophage K1F, a T7-like phage that has acquired the ability to replicate on K1 strains of *Escherichia coli*. J. Bacteriol. *187*, 8499–8503.

Scholl, D., Rogers, S., Adhya, S., and Merril, C.R. (2001). Bacteriophage K1-5 encodes two different tail fiber proteins, allowing it to infect and replicate on both K1 and K5 strains of *Escherichia coli*. J. Virol. *75*, 2509–2515.

Schroder, G., and Lanka, E. (2005). The mating pair formation system of conjugative plasmids-A versatile secretion machinery for transfer of proteins and DNA. Plasmid *54*, 1–25.

Scott, A.E., Timms, A.R., Connerton, P.L., Loc Carrillo, C., Adzfa Radzum, K., and Connerton, I.F. (2007). Genome dynamics of *Campylobacter jejuni* in response to bacteriophage predation. PLoS Pathog. *3*, e119.

Scott, J.R., and Churchward, G.G. (1995). Conjugative transposition. Annu. Rev. Microbiol. *49*, 367–397.

Sharp, P.M., and Li, W.H. (1986). An evolutionary perspective on synonymous codon usage in unicellular organisms. J. Mol. Evol. *24*, 28–38.

Shen, P., and Huang, H.V. (1986). Homologous recombination in *Escherichia coli*: dependence on substrate length and homology. Genetics *112*, 441–457.

Shen, P., and Huang, H.V. (1989). Effect of base pair mismatches on recombination via the RecBCD pathway. Mol. Gen. Genet. *218*, 358–360.

Siddique, A., and Figurski, D.H. (2002). The active partition gene *incC* of IncP plasmids is required for stable maintenance in a broad range of hosts. J. Bacteriol. *184*, 1788–1793.

Simpson, D.J., Dawson, L.F., Fry, J.C., Rogers, H.J., and Day, M.J. (2007). Influence of flanking homology and insert size on the transformation frequency of *Acinetobacter baylyi* BD413. Environ. Biosafety Res. *6*, 55–69.

Smeets, L.C., and Kusters, J.G. (2002). Natural transformation in *Helicobacter pylori*: DNA transport in an unexpected way. Trends Microbiol. *10*, 159–162; discussion 162.

Sode, K., Tatara, M., Takeyama, H., Burgess, J.G., and Matsunaga, T. (1992). Conjugative gene transfer in marine cyanobacteria: *Synechococcus* sp., *Synechocystis* sp. and *Pseudanabaena* sp. Appl. Microbiol. Biotechnol. *37*, 369–373.

del Solar, G., Alonso, J.C., Espinosa, M., and Diaz-Orejas, R. (1996). Broad-host-range plasmid replication: an open question. Mol. Microbiol. *21*, 661–666.

Stedman, K.M., She, Q., Phan, H., Holz, I., Singh, H., Prangishvili, D., Garrett, R., and Zillig, W. (2000). pING family of conjugative plasmids from the extremely thermophilic archaeon *Sulfolobus islandicus*: insights into recombination and conjugation in Crenarchaeota. J. Bacteriol. *182*, 7014–7020.

Sullivan, M.B., Waterbury, J.B., and Chisholm, S.W. (2003). Cyanophages infecting the oceanic cyanobacterium *Prochlorococcus*. Nature *424*, 1047–1051.

Sung, J.M., and Lindsay, J.A. (2007). *Staphylococcus aureus* strains that are hypersusceptible to resistance gene transfer from enterococci. Antimicrob. Agents Chemother. *51*, 2189–2191.

Sutherland, I.W., Hughes, K.A., Skillman, L.C., and Tait, K. (2004). The interaction of phage and biofilms. FEMS Microbiol. Lett. *232*, 1–6.

Taddei, F., Matic, I., and Radman, M. (1995). cAMP-dependent SOS induction and mutagenesis in resting bacterial populations. Proc. Natl. Acad. Sci. U.S.A. *92*, 11736–11740.

Taylor, A., and Smith, G.R. (1980). Unwinding and rewinding of DNA by the RecBC enzyme. Cell *22*, 447–457.

Terns, M.P., and Terns, R.M. (2011). CRISPR-based adaptive immune systems. Curr. Opin. Microbiol. *14*, 321–327.

Tock, M.R., and Dryden, D.T. (2005). The biology of restriction and anti-restriction. Curr. Opin. Microbiol. *8*, 466–472.

Thomas, C.M., and Nielsen, K.M. (2005). Mechanisms of, and barriers to, horizontal gene transfer between bacteria. Nat. Rev. Microbiol. *3*, 711–721.

Tuller, T. (2011). Codon bias, tRNA pools and horizontal gene transfer. Mob. Genet. Elements *1*, 75–77.

Van Ham, R.C., González-Candelas, F., Silva, F.J., Sabater, B., Moya, A., and Latorre, A. (2000). Postsymbiotic plasmid acquisition and evolution of the repA1-replicon in *Buchnera aphidicola*. Proc. Natl. Acad. Sci. U.S.A. *97*, 10855–10860.

de Vries, J., and Wackernagel, W. (2002). Integration of foreign DNA during natural transformation of *Acinetobacter* sp. by homology-facilitated illegitimate recombination. Proc. Natl. Acad. Sci. U.S.A. *99*, 2094–2099.

Wang, W., Swevers, L., and Iatrou, K. (2000). *Mariner* (*Mos1*) transposase and genomic integration of foreign gene sequences in *Bombyx mori* cells. Insect Mol. Biol. *9*, 145–155.

Weaver, K.E., Kwong, S.M., Firth, N., and Francia, M.V. (2009). The RepA_N replicons of Gram-positive bacteria: a family of broadly distributed but narrow host range plasmids. Plasmid *61*, 94–109.

Weinbauer, M.G. (2004). Ecology of prokaryotic viruses. FEMS Microbiol. Rev. *28*, 127–181.

Westra, E.R., Pul, U., Heidrich, N., Jore, M.M., Lundgren, M., Stratmann, T., Wurm, R., Raine, A., Mescher, M., Van Heereveld, L., *et al.* (2010). H-NS-mediated repression of CRISPR-based immunity in *Escherichia coli* K12 can be relieved by the transcription activator LeuO. Mol. Microbiol. *77*, 1380–1393.

Worth, L., Jr., Clark, S., Radman, M., and Modrich, P. (1994). Mismatch repair proteins MutS and MutL inhibit RecA-catalyzed strand transfer between diverged DNAs. Proc. Natl. Acad. Sci. U.S.A. *91*, 3238–3241.

Zahrt, T.C., Mora, G.C., and Maloy, S. (1994). Inactivation of mismatch repair overcomes the barrier to transduction between *Salmonella typhimurium* and *Salmonella typhi*. J. Bacteriol. *176*, 1527–1529.

The Evolution of Horizontally Transferred Genes: A Model for Prokaryotes

5

Iñaki Comas and Fernando González-Candelas

Abstract

Horizontal gene transfer is a pervasive evolutionary process which has developed and is still developing an essential role in shaping biodiversity through providing opportunities for innovation, moving determinants of functions among taxa, opening opportunities for colonization of new niches, or acting as a catalyst for adaptation. However, its importance in evolution has only recently begun to be recognized. Reasons for this relatively late recognition of HGT and its relevance stem from two main sources. One is the availability, or rather the lack thereof, of appropriate information to infer the existence of HGT. This shortcoming has now been widely overcome with the large, and still growing at a very fast rate, number of completely sequenced genomes, which allow for precise phylogenetic reconstructions, along with the deployment of theoretical models necessary for inferring evolutionary events in deep-time. The second source is the appreciation that HGT does not leave an indelible stamp on the transferred genes because these continue evolving under a different genomic, and often ecological, environment which usually act synergistically to erase the initially clear marks that go along with those genes. In this chapter we provide an overview of these processes along with a model proposal that will help to better understand the consequences of the continuous evolution of laterally transferred genes at different evolutionary time-scales.

A model from Xanthomonadales and *Legionella pneumophila*

Xanthomonadales are plant pathogens included in the gamma-Proteobacteria division. One relevant feature of Xanthomonadales is that their genomes are the result of multiple events of HGT from other Proteobacteria to the ancestral genome of the group (Fig. 5.1A) (Comas *et al.*, 2006). These results and recent findings by other groups have prompted us to delineate a model for the effects of lateral gene transfer along time on gene contents and phylogenomic analyses of genomes involved in HGT. This model is expected to hold for highly promiscuous bacteria where the main constraints to transfers are those derived from incompatibilities in genomic architecture and, therefore, from the time since divergence of the two lineages considered. This results in web-like structures for the phylogenies of HGT-prone taxa, such as Xanthomonadales, apparently due to absence of selective pressure against the majority of transfers, thus representing the most extreme case of genetic promiscuity among bacteria. This almost free-exchange scenario is represented in Fig. 5.2A.

The model developed for Xanthomonadales and similar taxa in terms of HGT is based on two assumptions:

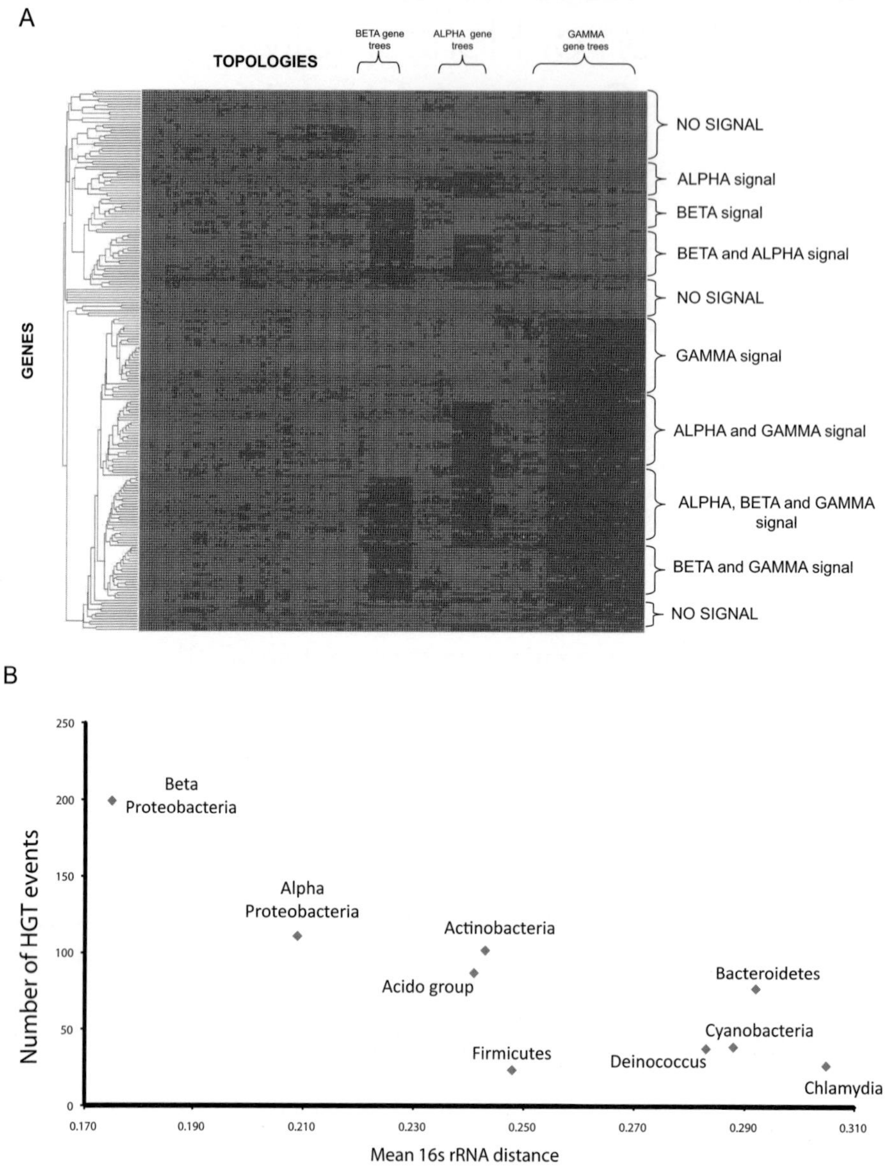

Figure 5.1 Likelihood of horizontal gene transfer decays with phylogenetic distance. (A) To differentiate between similar gene tree topologies we computed a tree likelihood comparison statistic that allows discerning between statistically different topologies. The heat-map shown in the figure represents an unsupervised clustering of gene trees and gene alignments according to whether the latter reject (grey dots) or not other gene trees topologies (black dots). Islands of similar gene tree topologies are discovered in this way. As shown in the figure some gene trees place Xanthomonadales as Gammaproteobacteria while others shows a beta or alpha signal, revealing a mosaic genome created by a large number of horizontal gene transfer events. (B) Decay in the number of horizontal gene transfer events detected after analysing the pan-genome of *Legionella pneumophila*. While decay in intra-specific rates of recombination has been known for a very long time, this is the first time that an analogue phenomenon but at much broader phylogenetic distances is proven.

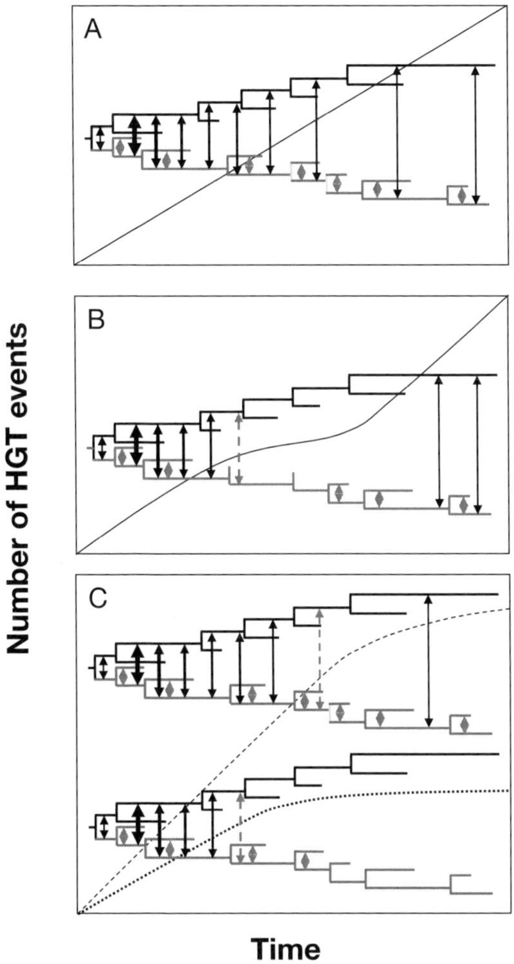

Figure 5.2 Changes in the frequency of HGT under different theoretical ecological scenarios illustrated by real examples. (A) The scenario for highly promiscuous bacteria like Xanthomonadales. A continuous import of foreign DNA happened during evolutionary history. This is a free-exchange scenario that is common to other groups and in which gene loss is expected to keep genome sizes on the range observed for bacteria. Many of these gene losses can be the result of natural selection acting on the newly acquired genes. This scenario is likely to happen in free-living or opportunistic bacteria with plenty of opportunities for gene exchange. (B) The scenario for a bacterial lineage in which the income of transferred genes is reduced as it adapts to a new niche thanks to previously acquired genomic islands. Once established, the new lineage starts to diversify and to receive again foreign genetic material. This case is exemplified by the diversification of *Phroclorococcus* strains in ecotypes that differ in the composition and distribution of several genomic islands. (C) The scenario for endosymbionts and some intracellular pathogens (for example *Mycobacterium leprae*). In this case the acquisition of genetic material can result in the opportunity to invade a new host, eventually as a pathogen. Endosymbionts and intracellular pathogens are an extreme example for this case, as their intracellular habitat prevents recombination events and transfer of genes in the majority of cases (dotted line). Non-intracellular genomes retain the capacity to accept foreign genetic material but usually with fewer opportunities (dashed line). The arrows within genealogies represent likelihood of transfers between and within nascent grey and black lineages over time, with likelihoods represented by the thickness of the arrows.

1 Genetic exchange in bacteria is widespread but not random, being more likely between closely related taxa than between evolutionarily distant groups (Gogarten *et al.*, 2002).
2 The consequences of gene transfers on genome phylogenies differ depending on the time elapsed since the transfer and from the time of divergence between the donor and the recipient.

If both principles are true then it is expected that the present preferential partners of Xanthomonadales are other genomes of the same clade, whereas in the past, when Proteobacteria and Xanthomonadales were nascent lineages, the preferential partners were other Proteobacteria, whose extant descendants may now be highly divergent from the Xanthomonadales. Our analysis of Xanthomonadales genomes corroborates this prediction.

Genome-wide phylogenies show two relevant facts: on the one hand, the Xanthomonadales clade is monophyletic for all the gene trees, thus showing that every gene analysed was already present in the last common ancestor of this group. On the other hand, a consensus tree failed to identify a single, common origin for these genes (Comas *et al.*, 2006). The second observation clearly points to ancient HGT to the ancestor of the genomes from other Proteobacteria, a possibility corroborated in further analyses. The first observation is indicating either that vertical inheritance was the dominant force during the diversification of the clade or that there are HGT events not detected by phylogenetic analyses at the genome-level whose signal is also vertical-like. An analysis of the composition of the *Xanthomonas citri* genome revealed the presence of a significant fraction of atypical genes (22%). Most of them were not detected in our phylogenetic analyses, thus corroborating that recent intra-clade exchanges exist and, therefore, that the cohesion of the clade observed in the summaries of gene trees is fuelled not only by vertical descent but also by horizontal gene transfer as predicted elsewhere (Gogarten *et al.*, 2002).

Let us now consider another model organism we have been working on in our laboratory. *Legionella pneumophila* is the causative agent of legionnaires' disease. It can be found in aquatic environments, free-living or in biofilms, or infecting human macrophages. One remarkable feature of *Legionella* is that it can survive also within amoebas, sharing this niche with many other bacterial species (Coscollá *et al.*, 2011). In fact, the amoeba host has been proposed as a training field for many human pathogens able to invade cells (Molmeret *et al.*, 2005). We have recently performed a phylogenomic analysis of the core genome of four *Legionella pneumophila* strains in search of possible HGT events. This analysis revealed that *Legionella* genomes harbour a large fraction (about 42% of 1700 genes included in the study) of seemingly horizontally transferred genes from different species, most of them with close relatives infecting amoebas. Furthermore, despite these increased chances for HGT deriving from the shared intracellular environment (see also Chapter 4), we were able to demonstrate that the number of HGT events in the *Legionella* genomes decreases with increasing phylogenetic distances from the putative donor group (Fig. 5.1B).

Therefore both the Xanthomonadales and the *Legionella pneumophila* genomes point towards the two main factors regulating the amount of HGT between two species: the phylogenetic distance between them (i.e. intrinsic factors) and the ecological opportunity for the exchange (i.e. extrinsic factors).

Non-random genetic exchanges: internal and external factors

As previously mentioned, there are different factors limiting a widespread, random exchange of genetic material among bacteria. Most of these factors are known but those belonging to the genome architecture/compatibility class have been reported only recently.

Recombination in bacterial populations is known to decrease with divergence. In fact, this is a by-product of the mechanisms involved in the process. The mismatch-correction systems of bacteria are capable of distinguishing foreign DNA based on its genetic divergence from the recipient (Majewski and Cohan, 1998; Cohan, 2002). Therefore, this mechanism represents a barrier to homologous recombination at the population level and, in consequence, it delineates a first boundary to gene exchange among lineages.

For years, bacteria were considered as clonal both at the population and the species levels (Orskov and Orskov, 1983; Stenderup and Orskov, 1983). Multilocus sequence typing analyses of different bacterial populations and the sequencing of multiple strains of the same species are aiding to study the role of recombination in bacterial evolution at the population level (Maiden, 2006). These analyses are revealing that most bacterial populations are far from being clonal and that recombination events can be detected by *in vitro* and *in silico* analyses (Perez-Losada *et al.*, 2006). The role of recombination in generating genetic variation and its contribution to overall genome plasticity is very different from one species to another. Estimates range from species in which recombination seems to be the major force of adaptive evolution (e.g. *Neisseria meningitidis*, *Streptococcus pneumoniae*, *Staphylococcus aureus*) to those that seem to evolve mainly by point mutations or other mechanisms, with recombination as a residual process (e.g. *Mycobacterium tuberculosis*, *Yersinia pestis*, *Bacillus anthracis*) (Feil and Spratt, 2001). However, most inferences on bacterial population structure are being reconsidered in the light of current analytical techniques that, as MLST, are changing dramatically long-held views such as the clonality of *Escherichia coli* populations (Wirth *et al.*, 2006). Nevertheless, these analyses have shown not only differences among taxa but also important differences within the corresponding genomes. Genes seem to have different recombination rates and, on top of that, highly recombining regions alternate with low recombining regions in bacterial genomes.

Nascent lineages that cannot (legitimately) recombine because of the accumulated divergence can acquire DNA from illegitimate sources. Instead of homologous recombination, other mechanisms can help to exchange genetic material among distant bacteria (Thomas and Nielsen, 2005). However, they are partly out of the scope of this review as we are interested in (1) reviewing mechanisms that prevent HGT between lineages, and (2) showing that the by-product of these mechanisms is the preferential sharing of DNA between evolutionarily close taxa.

Internal factors

Genomic architecture

Progress in the analyses of the structure of bacterial genomes, driven by the continuous sequencing of new genomes, is allowing their description and characterization beyond the cataloguing of genes included therein (Rocha, 2004). For example, the preferential distribution of genes in the leading strand has been usually explained by a selective pressure to avoid the collision of machineries in the leading strand during the simultaneous replication and transcription processes in bacteria. This would allow a better replication and higher expression of genes (Brewer, 1988) but essentiality seems to be a more important factor, even constraining the number of possible rearrangements (Rocha and Danchin, 2003a,b). Recent studies have also identified codon usage domains in bacterial chromosomes that are important for the control of gene expression (Bailly-Bechet *et al.*, 2006). These results support the idea of domains in bacterial chromosomes larger than genes, operons or even 'über-operons' (Boccard *et al.*, 2005). The alteration of these macrodomains by rearrangements or the insertion of foreign sequences could have a detrimental effect on the fitness of the organism and, therefore, a 'compatibility filter' must operate in order to maintain chromosomal properties.

Such a filter has already been described by Lawrence and collaborators (Lawrence and Hendrickson, 2004; Hendrickson and Lawrence, 2006). These authors analysed the distribution of a certain kind of octamers along the bacterial chromosome. Repeats of these sequences are usually found in bacterial genomes in both strands; however, in some cases they are distributed asymmetrically and skewed towards the terminus of replication. This characteristic distribution seems to be associated to a specific role in the replication of bacteria; they are recognized by protein FtsK, whose function is related to translocation of the chromosome during cell division. The protein uses the accumulation of this repeat sequence as a signal for the location of the terminus of replication. Therefore, for a correct cell division it is necessary to maintain the skewed and asymmetrical distribution of these octamers. This possibility of selection at the level of chromosome structure could have important implications in bacterial chromosome dynamics. For instance, only inversions including the origin or terminus of replication might be evolutionarily feasible because other alternatives imply a disruption in the accumulation of these repetitive sequences and, in consequence, a disruption in the action of FtsK. Of more relevance in this context, DNA fragments compatible with the distribution of these repetitive units in a recipient genome will have more chances of being successfully transferred. Until now, different octamer sequences with similar distribution to the one described above have been identified and are usually shared by related taxa. This implies that successful transfers will be more likely between close, and therefore compatible, genomes.

Molecular mechanisms

The best-known mechanism to control the uptake of DNA in Bacteria is the methylation of a genome's own genetic material. DNA molecules without the specific methylation pattern are digested (Jeltsch, 2003). Recently one of these systems has been proposed as being responsible for foreign DNA control in *Staphylococcus aureus* (Waldron and Lindsay, 2006). These type I restriction-methylation systems require the action of three genes for the methylation process (*hsdR, hsdM, hsdS*). Variations in these genes throughout sequenced *S. aureus* genomes and in multistrain microarray analyses showed a significant correlation with the ten known dominant lineages of this species. Furthermore, the recognition of a sequence as alien by *S. aureus* lineages is based on different sequence profiles of these genes. The effect of this biased methylation is that (1) intra-lineage exchanges are more likely than inter-lineage exchanges and (2) there are preferential sharing links between different lineages. Obviously, this mechanism also acts upon sequences obtained from other species and therefore its presence results in a more likely exchange between *S. aureus* strains to the point that it seems to be responsible for the identification of different lineages and the clonal structure of the species.

Another largely unresolved question about bacterial genome architecture is the characteristic GC/AT ratio (Bentley and Parkhill, 2004). Sometimes, lower GC contents are clearly associated with certain lifestyles, mainly in obligate intracellular pathogens and endosymbionts (Moran and Wernegreen, 2000). However, among the remaining genomes differences in GC content are remarkable and genus-specific. A possible role for these biases has been proposed. The protein H-NS has the ability to bind DNA, acting as a transcriptional regulator. Two independent experiments were able to determine that an important fraction of genes targeted by H-NS were low-GC compared with the average genome (Lucchini *et al.*, 2006; Navarre *et al.*, 2006). A low GC content together with the absence of these genes from the *Salmonella* core genome suggest that H-NS can act on HGT genes and repress their expression in a process called 'xenogenic' silencing. The silencing of these genes prevents their possible negative effects on the fitness of the genome. What is the fate of these silenced genes? Apparently bacteria have evolved mechanisms in order to selectively activate some of these genes (Navarre *et al.*, 2007). The fate of the remaining genes could be their pseudogenization. The H-NS mechanism in *Salmonella* suggests that transfers from donors with similar GC content maybe more likely to be retained in the genome because the H-NS mechanism is not at work. H-NS is a general regulator in bacteria with homologues mainly found in Gram-negative bacteria suggesting that silencing of recently acquired genes can be a general mechanism to control widespread horizontal gene transfer events (Tendeng and Bertin, 2003).

Functional association

Since the publication of the complexity hypothesis by Jain *et al.* (1999) it has been accepted that genes belonging to informational categories are less prone to HGT than operational ones. The hypothesis is based on two observations: (1) HGT has been continuous in the evolutionary history of bacteria, and (2) there are important differences between the degree of phylogenetic incongruence found in operational and metabolism-related genes. Jain *et al.* (1999) proposed that informational genes tend to be part of essential complexes interacting with many products whereas operational genes usually take part in simpler metabolic networks. The likelihood of integration and its effect on fitness should be very different, thus resulting in more successful transfers of operational genes than informational ones. Posterior studies with more genes and genomes have shed light on the importance of the distinction among operational and informational genes. Although there is clearly an association between the two factors it seems that the pattern is more complex than initially postulated and it is not only a by-product of the connectivity of the transferred gene.

In support of the complexity hypothesis, Nakamura *et al.* (2004) identified metabolism-related genes (cell surface, DNA binding and pathogenicity-related functions) as the more frequently transferred group and found very few cases of transfers involving informational genes. In addition, a more recent study with a very different approximation corroborated these observations (Pal *et al.*, 2005). These authors studied the influence of different factors on the stability of metabolic networks of bacteria. They identified several instances of horizontal gene transfer events and mapped them onto the metabolic network of *E. coli*. Their analyses revealed that (1) most genes involved in HGT belong to metabolism-related categories, and (2) these genes usually occupy external nodes of the metabolic networks. These observations supported the complexity hypothesis because it is expected that genes attached to external nodes of a network carry out non-essential functions, which, in turn, are mainly encoded by operational genes. Therefore, the numbers of interactions in which a newly acquired gene is involved are fewer than those interactions needed when the transferred gene corresponds to a central node of the network.

However, other studies analysing the phenomenon from a phylogenetic point of view only agree partially with these observations. Phylogenetic analysis of incongruence has shown the presence of possible transfer events in an important proportion of core gene datasets (Bapteste *et al.*, 2005; Susko *et al.*, 2006). Most of the genes in these cores are included in informational categories. Also, recombination events in the elongation factor complex, even at large phylogenetic distances, have been reported (Inagaki *et al.*, 2006). A phylogenomic approach method called 'embedded quartets' has allowed inference of the number of transfer events among Cyanobacteria genomes and between Cyanobacteria and external species (Zhaxybayeva *et al.*, 2006). The analysis revealed these genomes as mosaics and the analysis of functional categories showed two distinct patterns. Transfers among cyanobacterial genomes showed no clear pattern, whereas those transfers that implied a non-cyanobacterial partner seemed to be biased towards operational genes, thus implying that transfers of informational genes are more likely between closely related genomes. The same pattern was observed in the screening of Xanthomonadales genomes in search of transfer events to the ancestor of the group and within the group (Comas *et al.*, 2006). Most ancestral transfers were from other Proteobacteria and showed no clear pattern of functional association although their mosaicism at a functional and genome scale prevented assigning a single phylogenetic origin to the group.

More recently the complexity hypothesis has been revisited and reformulated in light of new experimental data. The number of protein interactions seems to be a better predictor of the likelihood of transfer of a gene than the functional category it belongs to (Cohen *et al.*, 2011), a hypothesis supported by some experimental models of evolution (Omer *et al.*, 2010).

External factors

Ecological opportunity

The last stage in the genomic revolution, started with the publication of the *H. influenzae* genome in 1995 (Fleischman *et al.*, 1995), has been the study of microbial communities using metagenomic approaches. These methodologies are aimed at identifying not only the bacterial diversity in a specific niche but also the gene composition of its bacterial communities and the metabolic capabilities derived from them. There are two main factors to be considered in order to achieve these objectives. On the one hand, laboratory techniques have been developed that allow the fast isolation and sequencing of samples and on the other hand it is necessary to have a large enough taxon sampling of complete microbial genomes in order to assign sequences to their most closely related taxa.

In the last years, many studies have analysed the gene and species composition of very different environments. Some of them have revealed a low bacterial diversity with communities dominated by two or a few bacterial taxa, whereas other studies have identified niches in which bacterial diversity is much higher, with complex ecological networks relating them.

The number of new genes, those with no known function or not identified in any previously sequenced genome, is extraordinarily high even in communities composed by bacteria with clear relatives in the databases. All these studies are highlighting the vast amount of unknown microbial diversity and the limitations that this ignorance is imposing to our current views of bacterial evolution.

The analysis of environmental metagenomes is important in the context of HGT because one of the most important extrinsic factors affecting the process is ecological opportunity (see Chapter 4 for more details). Two genomes may be compatible in terms of genome architecture but if they occupy very different niches with no opportunity for contact then successful HGT events would be very unlikely. For example, Tyson *et al.* (2004) analysed the bacterial community structure of samples recovered from acid mine drainage (AMD) microbial biofilms. Only a few lineages dominated the community, mostly *Leptospirillum* species. The low-complexity of the environment seems to explain the presence of this dominant lineage and other low frequency Bacteria and Archaea (the *Ferroplasma* type II genome was sequenced from this study). In such an environment, the routes for DNA transfer are constrained. On the one hand, recombination seems to be the major evolutionary force driving speciation of the dominant *Leptospirillum* genomes. On the other hand, a few more cases of ancient HGT could be inferred including cases of inter-domain exchange. At the other extreme, the analysis of samples from the Sargasso Sea revealed a richer ecological structure composed by a range of 1800 to 48,000 genomic species (Venter *et al.*, 2004). Although almost all bacterial phyla were represented, most of the species proved to be related to already sequenced Proteobacteria genomes. Another remarkable fact was the presence of spatial patchiness in the frequency and distribution of these species and the identification of around 1.2 million new genes. Therefore, in such a complex community ecological opportunity for genetic exchange at almost all phylogenetic distances is possible in contrast to the AMD environment.

The two examples outlined above highlight also the need for new conceptual frameworks for the evolutionary genomic analysis of bacteria. The term 'pan-genome' has been proposed to describe all the genes present in any of the genomes from a species (Tettelin *et al.*, 2005). Species such as *Streptococcus agalactiae* have an open pan-genome and therefore it is expected that even the sequencing of 100 more strains will add new genes to it (Tettelin *et al.*, 2005). The analysis of closely related *Shigella/Escherichia* species shows similar results (Chen *et al.*, 2006). Other genera, like *Bacillus,* seem to have a closed pan-genome: almost all genes that are part of one of its genomes are also present in the rest and the sequencing of further genomes will probably not add much to the known catalogue (Tettelin *et al.*, 2005). Analogously, a pan-genome for metagenomic studies could be outlining the so-called microbiome. An open pan-genome/microbiome clearly points to HGT as the main agent of evolutionary novelty. Focusing in the two metagenome examples given above, an environment such as the AMD dominated by *Leptospirillum* species is expected to have a closed microbiome while an environment such as that of the Sargasso Sea would have an unapproachable microbiome. Therefore, the opportunity for transfers is very different in these two environments, being much higher in the second case. It is also important to note that transfers across very large distances are not impossible as demonstrated by Tyson *et al.* (2004). However, as illustrated in this study, transfers from Bacteria to Archaea are usually related with adaptation to very specific niches, in which case the transfer of functions allowing survival in this niche is associated to a high fitness.

Phage genomics

Phage-mediated transfers are one of the possible mechanisms introducing foreign DNA in a bacterial genome. In the context of mechanisms and limitations for HGT, phages can be analysed from different points of view. They can be used as model organisms to study characteristics expected to be present in more complex organisms like Bacteria. Alternatively, they can be studied attending to their influence on bacterial chromosome evolution, as carriers of foreign genetic material.

Phages are clearly mosaic genomes resulting from multiple homologous and non-homologous recombination events. This mosaicism has been described both for double- and single-strand DNA phages (Lawrence *et al.*, 2002). However, the analysis of different T4-type phages has revealed limitations to rampant gene transfer (Filee *et al.*, 2006). These authors located two regions that seem not to be affected by HGT and whose main feature is the conservation of synteny across the analysed genomes. The genes composing these regions have coupled metabolic functions, characterized by many interactions. A disruption of these regions could result in important losses of fitness. As we have described, the conservation of genomic architecture seems to be an important factor for successful transfers in bacterial genomes. The evidence derived from phage genomics corroborates this importance.

Our current view of phages has changed from an early view as agents for the transportation of antibiotic resistance genes to consider them as catalysts of bacterial evolution (Canchaya *et al.*, 2003). Their lysogenic action used to be considered solely as the cause of conversion from a non-pathogen strain to a pathogen organism as shown by the phage encoded toxins of the causative agents of diphtheria or shigellosis (Brussow *et al.*, 2004). However, the ecology of phages also has to be taken into account. They have limited host ranges, mainly at the species level, although cases of broader ranges covering large phylogenetic distances have been described (Jensen *et al.*, 1998; Beumer and Robinson, 2005). It is known that bacterial strains usually share a common phage pool and therefore phage-mediated transfers are more likely between more closely related strains. The same example of the acid mine drainage used above is also useful to illustrate this point. The authors demonstrated that recombination between *Leptospirillum* species and between *Ferroplasma* species are phage dependent, as revealed by the analysis of their genomes (Tyson *et al.*, 2004). In another study, the analysis of *Prochlorococcus* genomes through a wide range of ocean light and depth conditions revealed that there are different ecotypes depending of these variables (Coleman *et al.*, 2006). The distinguishing feature of these ecotypes is the composition of different genomic islands in their genomes that have been introduced most likely by phages. These studies are pointing to phages as the main via of HGT in closely related microbial genomes as revealed by multistrain analyses of sequenced genomes (Brussow *et al.*, 2004; Aziz *et al.*, 2005). Finally, there is a dual role for phages in the transfer of genetic material. There are cases of phages with very restricted host ranges combined with cases of phages with a wide host spectrum and cosmopolitan distribution, which implies that they can act as a link between bacterial species and also across large geographic areas.

A proposal for Bacteria and Archaea

Our proposal for the nature and impact of gene transfer processes on microbial genomes is based on the idea that the likelihood of transfers decays with the phylogenetic distance between the lineages involved, in analogy to the lower recombination rate between more divergent sequences. We have explained this effect as a by-product of different mechanisms that allow bacteria to prevent the incorporation, silence or eliminate foreign material from more or less distant sources. Some of these mechanisms help to define intra-species lineages whereas others prevent inter-species exchange and, presumably, other mechanisms that surely act at other phylogenetic depths have still to be identified. However, horizontal gene transfer events between distant taxa are still detectable. Among them only those transfers from distant sources with a high, positive effect on the fitness of the recipient will be retained. Usually, those successful transfers are dependent on environmental conditions like the exchange of niche-specific genes between Bacteria and Archaea that share the same extreme environments (Mongodin *et al.*, 2005).

Although the decay of recombination rates with divergence has been demonstrated experimentally (Majewski and Cohan, 1998), only now the growing number of sequenced genomes is allowing to study the age and number of transfers between species. There is a growing body of evidence supporting the assumption that the likelihood of horizontal gene transfers decays with time since divergence from the common ancestor (Andam and Gogarten, 2011; Popa *et al.*, 2011). As mentioned previously, the evolution of Xanthomonadales has been driven by ancient and recent HGT events from other Proteobacteria and from other Xanthomonadales genomes respectively (Comas *et al.*, 2006). We tested more properly the hypothesis of the decay of HGT events with distance using the genome of the opportunistic pathogen *Legionella pneumophila*. As a difference with other genome-wide studies of horizontal gene transfer events we decided to focus on an single species and tested its likelihood of receiving genes from other bacterial groups after removal of those gene families with poor phylogenetic signal. The approach allowed us, firstly, to test for the number of events occurred between *Legionella* and other groups of bacteria and, secondly, to differentiate between noise and signal when doing phylogenetic assignments. In that way we have shown for the first time in Bacteria a high correlation between phylogenetic distance and the number of horizontal gene transfer events to a specific genome (Coscollá *et al.*, 2011). Although not explicitly tested as in our work with *L. pneumophila*, other studies pointing towards similar results have been published. The Cyanobacteria genomes analysed by Zhaxybayeva *et al.* (2006) revealed that around 50% of their genes have suffered at least one HGT event. What is more relevant for our proposal, intra-Cyanobacteria transfers were more frequent than between Cyanobacteria and external genomes. Other studies have shown similar patterns in Corynebacteria (Marri *et al.*, 2007). In addition, analyses of multiple genomes covering large phylogenetic distances point towards the hypothesis of phylogenetic preferential sharing. In the most complete analysis published up to now, Beiko and collaborators (2005) screened 144 sequenced genomes including Bacteria and Archaea. They derived a supertree from the single-gene phylogenies obtained from all the gene families retrieved from these genomes and then tried to reconcile every gene tree and the supertree. As the reconciliation of a gene tree equals to a putative HGT event, their conclusion was that most of the inferred transfers were between genomes of the same group, for example between genomes of Gammaproteobacteria. Therefore the analysis showed as preferential partners in HGT events those of closely related genomes, as predicted by Gogarten *et al.* (2002).

To simplify our argument we will consider a genome whose capability to accept new genetic material is constant though time. In this case, a linear income of foreign material during the evolutionary history of the genome is expected, as shown in Fig. 5.2. Therefore, the current genome derived from this lineage will present ancient, recent and ongoing

HGTs. From a methodological point of view the detection of these events requires very different approximations.

Ongoing transfers are composed mainly by homologous recombination within the lineage. This kind of transfers is only detectable in a population genetics framework that allows inferring events and rates of recombination. There are many examples on how to use these techniques in bacterial populations (Perez-Losada *et al.*, 2006). From a phylogenomic point of view, there is no effect on genome phylogenies not only because there is not enough phylogenetic sampling but also because at this scale recombination acts synergistically with the vertical signal as a cohesive force of the group (Fraser *et al.*, 2007).

However, homologous recombination is not the only mechanism for acquiring foreign material. Recent transfers from other genomes can also be detected by surrogate methods (Lawrence and Ochman, 2002). These methods identify atypical regions in the genomes based on a significant difference in some compositional measure. Each genome is under different mutational pressures (Sueoka, 1993), which result in specific patterns of, for example, nucleotide composition (Lawrence and Ochman, 1997; Lawrence and Ochman, 1998; Ochman *et al.*, 2000), codon usage bias (Medigue *et al.*, 1991), dinucleotide frequencies (Karlin and Burge, 1995; Karlin, 2001) and sequence patterns detected by Markov models (Hayes and Borodovsky, 1998). In consequence, the introduction of new DNA by an HGT event results in the integration in the genome of a sequence with different features than those typical of the recipient genome (Ochman *et al.*, 2000). Usually each of these measures tends to generate different sets of atypical genes although it has been argued that the reason is that each one tests a different hypothesis, being best suited to detect transfers from different genomes (Azad and Lawrence, 2005). Although not all the atypical genes detected result from HGTs, they are a good measure of the recent and ongoing transfers in a genome. However, the main limitation of these methods is the reduction and eventual elimination of atypical signals with time (Lawrence and Ochman, 1997). This process, called amelioration, starts when the DNA integrates in the genome and therefore when the sequence starts to undergo the same biases than the rest of the genome. The process of amelioration is relatively fast and therefore the power to detect atypical genes decreases with time, being valid only for recent transfers. The main advantage of this approach is that it does not rely on a comparative analysis and therefore those genes excluded from a phylogenetic analysis owing to their presence in only one or a few lineages can be studied. Phylogenetic methods will detect these transfers if the divergence between the donor and receptor is large enough. For example, a recent transfer between Bacteria and Archaea is easy to detect whereas a recent transfer between *Salmonella typhimurium* and *Escherichia coli* will be more difficult and would be only detected if the number of strains per species used in the study is large enough to reveal the donor and recipient genomes.

Older transfers can be detected only by phylogenetic methods and their detection depends to a large extent on the time elapsed since the transfer. If the gene was transferred relatively recently, then it would show a clearly discordant phylogenetic signal, grouping the receptor in the donor's clade. However, if this transfer was older then the gene would carry two opposite signals: that from the donor, originated before the transfer, and the signal from the receptor, originated after the transfer. The first signal would suffer a process of phylogenetic amelioration, which is partly due to the compositional amelioration and to sharing selective pressures with related genomes. For example, positive selection could favour the same non-synonymous changes in the genes of a clade of genomes. Usually this translates into a lowly supported assignment of these genes to any clade. When the transfer is very ancient it is often impossible to differentiate between a transferred gene and the genes of the receptor genome because most of the phylogenetic history that could be inferred is shared with the rest of the related genomes. The deeper the transfer the more difficult it would be to detect it. This also explains the presence of noise in terms of horizontal signal. Those gene transfers that are old, but not very ancient, will show support both for the donor phylogeny and for the receptor phylogeny. Therefore, some noise, such that one is unable to differentiate between the assumed species tree and the transfer hypotheses, is the hallmark of an old transfer event and is revealing a mixture of donor and receptor signals.

According to this model, the lineage considered will be receiving foreign DNA at a more or less constant rate, depending on its environment, as shown in Fig. 5.2. Next, the interplay between selection and effective population sizes will act filtering each transfer event, which will lead to differential incorporation of foreign material over time. Particularly, selection will act so that newly acquired material will be incorporated either when it has almost no impact on fitness (neutral or quasi-neutral effect which will depend on the effective population size) or when it represents a strong fitness advantage as in the case of new niche occupation or acquisition of new capabilities to compete with related strains. A free-exchange scenario, in which a lineage has constant access to foreign material through time is an ideal situation that might fit highly promiscuous bacteria like Xanthomonadales, which seem to lack most selective filters that prevent gene exchange. Other bacteria could have different behaviours. For example, a free-living bacterial lineage could receive a gene island that eventually allows its adaptation to a new niche. During the adaptation process population sizes, recombination rates and possibilities of gene transfers usually reduce, thus lowering the transfer rate. However, as the lineage becomes adapted to the new environment, diversification by periodic selection episodes and the acquisition of niche-specific genes could lead to an increase in the number and diversity of bacterial lineages or ecotypes. These could differentiate up to a point in which recombination between them is prevented although non-homologous transfers continue (Fig. 5.2).

This seems to be the case for the distribution of *Prochlorococcus* ecotypes in the oceans (Johnson *et al.*, 2006). Despite sharing a core genome, these ecotypes present a very large variability in the remaining gene contents, which reside mostly in genomic islands. A high correlation between genomic island composition and structure is found with the niche each ecotype occupies. These islands have been acquired by phage-mediated transfers that, in consequence, acted as drivers of *Procholorococcus* diversification. The establishment of the lineage in a new niche allows new opportunities for recombination and gene sharing.

But the acquisition of genes that promote the invasion of a new niche could also imply that some new or old genetic material is no longer necessary because the organism has specialized in exploiting the new niche. This is the case for most pathogens derived from a free-living ancestor. The free-living genome receives transfers from other genomes. Eventually some of these transfers, presumably pathogenicity islands, allow it to infect a host as a pathogen. As the transition to a pathogenic lifestyle continues the capacity for receiving external genetic material decreases both because it is less important and because the environment offers fewer opportunities. Most bacterial pathogens show an intermediate pattern of reduced HGT capacity and genome decay signals due to constraints both in population sizes and in the selective advantages of new genetic material (Fig. 5.2). The most extreme cases are those of intracellular pathogens and endosymbionts, which couple specialization in the new host with loss of non-essential genes for their lifestyle. Intracellular pathogens in which these phenomena have been described are *Mycobacterium leprae* (Cole *et al.*, 2001) or *Ricketssia prowazeckii* (Müller and Martin, 1999) (Fig. 5.2). The genomes of bacterial endosymbionts undergo the same processes as obligatory intracellular pathogens but much more accentuated, with perfect gene-order conservation among strains and with no opportunity for gene exchange (Wernegreen, 2002). In these cases, horizontal gene transfers to the non-intracellular ancestors of these genomes might be detected but only if they have not been lost along the genome reduction process.

Conclusion

It has been estimated that every gene family has suffered at least one transfer event during its evolutionary history (Dagan and Martin, 2007). The impact of many aspects of horizontal gene transfer is still hotly debated (Ragan and Beiko, 2009) but currently it is clear that HGT is a major force shaping the genome contents of prokaryotes. From the initial reports showing the surprisingly high proportion of foreign genetic material in bacterial genomes we have now moved to quantitative approaches revealing the complex interplay between gene loss, horizontal gene transfer and prokaryotes' lifestyles. We think that the combination of shotgun sequences generated over the last 15 years with the large amount of information generated by the new sequencing technologies will give us insights into the horizontal gene transfer process to a scale and detail that will solve many open issues about its past, present and future impact on bacterial innovation.

References

Andam, C.P., and Gogarten, J.P. (2011). Biased gene transfer in microbial evolution. Nat. Rev. Micro. *9*, 543–555.

Azad, R.K., and Lawrence, J.G. (2005). Use of artificial genomes in assessing methods for atypical gene detection. PLoS Comput. Biol. *1*, e56.

Aziz, R.K., Edwards, R.A., Taylor, W.W., Low, D.E., McGeer, A., and Kotb, M. (2005). Mosaic prophages with horizontally acquired genes account for the emergence and diversification of the globally disseminated m1t1 clone of *Streptococcus pyogenes*. J. Bacteriol. *187*, 3311–3318.

Bailly-Bechet, M., Danchin, A., Iqbal, M., Marsili, M., and Vergassola, M. (2006). Codon usage domains over bacterial chromosomes. PLoS Comput. Biol.*2*, e37.

Bapteste, E., Susko, E., Leigh, J., MacLeod, D., Charlebois, R.L., and Doolittle, W.F. (2005). Do orthologous gene phylogenies really support tree-thinking? BMC Evol. Biol. *5*, 33.

Beiko, R.G., Harlow, T.J., and Ragan, M.A. (2005). Highways of gene sharing in prokaryotes. Proc. Natl. Acad. Sci. U.S.A. *102*, 14332–14337.

Bentley, S.D., and Parkhill, J. (2004). Comparative genomic structure of prokaryotes. Annu. Rev. Genet. *38*, 771–791.

Beumer, A., and Robinson, J.B. (2005). A broad-host-range, generalized transducing phage (SN-T) acquires 16S rRNA genes from different genera of bacteria. Appl. Environ. Microbiol. *71*, 8301–8304.

Boccard, F., Esnault, E., and Valens, M. (2005). Spatial arrangement and macrodomain organization of bacterial chromosomes. Mol. Microbiol. *57*, 9–16.

Brewer, B.J. (1988). When polymerases collide: Replication and the transcriptional organization of the *E. coli* chromosome. Cell *53*, 679–686.

Brussow, H., Canchaya, C., and Hardt, W.D. (2004). Phages and the evolution of bacterial pathogens: from genomic rearrangements to lysogenic conversion. Microbiol. Mol. Biol. Rev. *68*, 560–602.

Canchaya, C., Fournous, G., Chibani-Chennoufi, S., Dillmann, M.L., and Brussow, H. (2003). Phage as agents of lateral gene transfer. Curr. Opin. Microbiol. *6*, 417–424.

Chen, S.L., Hung, C.S., Xu, J., Reigstad, C.S., Magrini, V., Sabo, A., Blasiar, D., Bieri, T., Meyer, R.R., Ozersky, P., *et al.* (2006). Identification of genes subject to positive selection in uropathogenic strains of *Escherichia coli*: a comparative genomics approach. Proc. Natl. Acad. Sci. U.S.A. *103*, 5977–5982.

Cohan, F.M. (2002). Sexual isolation and speciation in bacteria. Genetica *116*, 359–370.

Cohen, O., Gophna, U., and Pupko, T. (2011). The complexity hypothesis revisited: connectivity rather than function constitutes a barrier to horizontal gene transfer. Mol. Biol. Evol. *28*, 1481–1489.

Cole, S.T., Eiglmeier, K., Parkhill, J., James, K.D., Thomson, N.R., Wheeler, P.R., Honore, N., Garnier, T., Churcher, C., Harris, D., *et al.* (2001). Massive gene decay in the leprosy bacillus. Nature *409*, 1007–1011.

Coleman, M.L., Sullivan, M.B., Martiny, A.C., Steglich, C., Barry, K., DeLong, E.F., and Chisholm, S.W. (2006). Genomic islands and the ecology and evolution of *Prochlorococcus*. Science *311*, 1768–1770.

Comas, I., Moya, A., Azad, R.K., Lawrence, J.G., and González-Candelas, F. (2006). The evolutionary origin of xanthomonadales genomes and the nature of the horizontal gene transfer process. Mol. Biol. Evol. *23*, 2049–2057.

Coscollá, M., Comas, I., and González-Candelas, F. (2011). Quantifying nonvertical inheritance in the evolution of *Legionella pneumophila*. Mol. Biol. Evol. *28*, 985–1001.

Dagan, T., and Martin, W. (2007). Ancestral genome sizes specify the minimum rate of lateral gene transfer during prokaryote evolution. Proc. Natl. Acad. Sci. U.S.A. 0606318104.

Feil, E.J., and Spratt, B.G. (2001). Recombination and the population structures of bacterial pathogens. Annu. Rev. Microbiol. *55*, 561–590.

Filee, J., Bapteste, E., Susko, E., and Krisch, H.M. (2006). A selective barrier to horizontal gene transfer in the T4-type bacteriophages that has preserved a core genome with the viral replication and structural genes. Mol. Biol. Evol. *23*, 1688–1696.

Fleischmann, R.D., Adams, M.D., White, O., Clayton, R.A., Kirkness, E.F., Kerlavage, A.R., Butt. C.J., Tomb, J.F., Dougherty, B.A., Merrick, J.M., *et al.* (1995). Whole-genome random sequencing and assembly of *Haemophilus influenzae* Rd. Science *269*, 496–512.

Fraser, C., Hanage, W.P., and Spratt, B.G. (2007). Recombination and the nature of bacterial speciation. Science *315*, 476–480.

Gogarten, J.P., Doolittle, W.F., and Lawrence, J.G. (2002). Prokaryotic evolution in light of gene transfer. Mol. Biol. Evol. *19*, 2226–2238.

Hayes, W.S., and Borodovsky, M. (1998). How to interpret an anonymous bacterial genome: machine learning approach to gene identification. Genome Res. *8*, 1154–1171.

Hendrickson, H., and Lawrence, J.G. (2006). Selection for chromosome architecture in bacteria. J. Mol. Evol. *62*, 615–629.

Inagaki, Y., Susko, E., and Roger, A.J. (2006). Recombination between elongation factor 1α genes from distantly related archaeal lineages. Proc. Natl. Acad. Sci. U.S.A. *103*, 4528–4533.

Jain, R., Rivera, M.C., and Lake, J.A. (1999). Horizontal gene transfer among genomes: The complexity hypothesis. Proc. Natl. Acad. Sci. U.S.A. *96*, 3801–3806.

Jeltsch, A. (2003). Maintenance of species identity and controlling speciation of bacteria: a new function for restriction/modification systems? Gene *317*, 13–16.

Jensen, E.C., Schrader, H.S., Rieland, B., Thompson, T.L., Lee, K.W., Nickerson, K.W., and Kokjohn, T.A. (1998). Prevalence of broad-host-range lytic bacteriophages of *Sphaerotilus natans*, *Escherichia coli*, and *Pseudomonas aeruginosa*. Appl. Environ. Microbiol. *64*, 575–580.

Karlin, S. (2001). Detecting anomalous gene clusters and pathogenicity islands in diverse bacterial genomes. Trends Microbiol. *9*, 335–343.

Karlin, S., and Burge, C. (1995). Dinucleotide relative abundance extremes: a genomic signature. Trends Genet.. *11*, 283–290.

Lawrence, J.G., and Hendrickson, H. (2004). Chromosome structure and constraints on lateral gene transfer. Dynam. Genet. 319–336.

Lawrence, J.G., and Ochman, H. (1997). Amelioration of bacterial genomes: rates of change and exchange. J. Mol. Evol. *44*, 383–397.

Lawrence, J.G., and Ochman, H. (1998). Molecular archaeology of the *Escherichia coli* genome. Proc. Natl. Acad. Sci. U.S.A. *95*, 9413–9417.

Lawrence, J.G., and Ochman, H. (2002). Reconciling the many faces of lateral gene transfer. Trends Microbiol. *10*, 1–4.

Lawrence, J.G., Hatfull, G.F., and Hendrix, R.W. (2002). Imbroglios of viral taxonomy: genetic exchange and failings of phenetic approaches. J. Bacteriol. *184*, 4891–4905.

Lucchini, S., Rowley, G., Goldberg, M.D., Hurd, D., Harrison, M., and Hinton, J.C.D. (2006). H-NS mediates the silencing of laterally acquired genes in Bacteria. PLoS Pathog. *2*, e81.

Maiden, M.C.J. (2006). Multilocus sequence typing of Bacteria. Annu. Rev. Microbiol. *60*.

Majewski, J., and Cohan, F.M. (1998). The effect of mismatch repair and heteroduplex formation on sexual isolation in *Bacillus*. Genetics *148*, 13–18.

Marri, P., Hao, W., and Golding, G.B. (2007). The role of laterally transferred genes in adaptive evolution. BMC Evol. Biol. *7*, S8.

Medigue, C., Rouxel, T., Vigier, P., Henaut, A., and Danchin, A. (1991). Evidence for horizontal gene transfer in *Escherichia coli* speciation. J. Mol. Biol. *222*, 851–856.

Molmeret, M., Horn, M., Wagner, M., Santic, M., and bu Kwaik, Y. (2005). Amoebae as training grounds for intracellular bacterial pathogens. Appl. Environ. Microbiol. *71*, 20–28.

Mongodin, E.F., Nelson, K.E., Daugherty, S., DeBoy, R.T., Wister, J., Khouri, H., Weidman, J., Walsh, D.A., Papke, R.T., Sanchez Perez, G., *et al.* (2005). The genome of *Salinibacter ruber*: Convergence and gene exchange among hyperhalophilic bacteria and archaea. Proc. Natl. Acad. Sci. U.S.A. *102*, 18147–18152.

Moran, N.A., and Wernegreen, J.J. (2000). Lifestyle evolution in symbiotic bacteria: insights from genomics. Trends Ecol. Evol. *15*, 321–326.

Müller, M., and Martin, W. (1999). The genome of *Rickettsia prowazekii* and some thoughts on the origin of mitochondria and hydrogenosomes. Bioessays *21*, 377–381.

Nakamura, Y., Itoh, T., Matsuda, H., and Gojobori, T. (2004). Biased biological functions of horizontally transferred genes in prokaryotic genomes. Nat. Genet. *36*, 760–766.

Navarre, W.W., Porwollik, S., Wang, Y., McClelland, M., Rosen, H., Libby, S.J., and Fang, F.C. (2006). Selective silencing of foreign DNA with low GC content by the H-NS protein in *Salmonella*. Science *313*, 236–238.

Navarre, W.W., McClelland, M., Libby, S.J., and Fang, F.C. (2007). Silencing of xenogeneic DNA by H-NSG-facilitation of lateral gene transfer in bacteria by a defense system that recognizes foreign DNA. Gene. Dev. *21*, 1456–1471.

Ochman, H., Lawrence, J.G., and Groisman, E.A. (2000). Lateral gene transfer and the nature of bacterial innovation. Nature *405*, 299–304.

Omer, S., Kovacs, A., Mazor, Y., and Gophna, U. (2010). Integration of a foreign gene into a native complex does not impair fitness in an experimental model of lateral gene transfer. Mol. Biol. Evol. *27*, 2441–2445.

Orskov, F., and Orskov, I. (1983). From the national institutes of health. Summary of a workshop on the clone concept in the epidemiology, taxonomy, and evolution of the Enterobacteriaceae and other bacteria. J. Infect. Dis. *148*, 346–357.

Pal, C., Papp, B., and Lercher, M.J. (2005). Adaptive evolution of bacterial metabolic networks by horizontal gene transfer. Nat. Genet. *37*, 1372–1375.

Perez-Losada, M., Browne, E.B., Madsen, A., Wirth, T., Viscidi, R.P., and Crandall, K.A. (2006). Population genetics of microbial pathogens estimated from multilocus sequence typing (MLST) data. Infect. Genet. Evol. *6*, 97–112.

Popa, O., Hazkani-Covo, E., Landan, G., Martin, W., and Dagan, T. (2011). Directed networks reveal genomic barriers and DNA repair bypasses to lateral gene transfer among prokaryotes. Genome Res. *21*, 599–609.

Ragan, M.A., and Beiko, R.G. (2009). Lateral genetic transfer: open issues. Phil. Trans. R. Soc. B *364*, 2241–2251.

Rocha, E.P.C. (2004). The replication-related organization of bacterial genomes. Microbiology *150*, 1609–1627.

Rocha, E.P.C., and Danchin, A. (2003a). Essentiality, not expressiveness, drives gene-strand bias in bacteria. Nat. Genet. *34*, 377–378.

Rocha, E.P.C., and Danchin, A. (2003b). Gene essentiality determines chromosome organisation in bacteria. Nucl. Acids. Res. *31*, 6570–6577.

Stenderup, J., and Orskov, F. (1983). The clonal nature of enteropathogenic *Escherichia coli* strains. J. Infect. Dis. *148*, 1019–1024.

Sueoka, N. (1993). Directional mutation pressure, mutator mutations, and dynamics of molecular evolution. J. Mol. Evol. *37*, 137–153.

Susko, E., Leigh, J., Doolittle, W.F., and Bapteste, E. (2006). Visualizing and assessing phylogenetic congruence of core gene sets: a case study of the gamma-proteobacteria. Mol. Biol. Evol. *23*, 1019–1030.

Tendeng, C., and Bertin, P.N. (2003). H-NS in Gram-negative bacteria: a family of multifaceted proteins. Trends Microbiol. *11*, 511–518.

Tettelin, H., Masignani, V., Cieslewicz, M.J., Donati, C., Medini, D., Ward, N.L., Angiuoli, S.V., Crabtree, J., Jones, A.L., Durkin, A.S., *et al.* (2005). Genome analysis of multiple pathogenic isolates of *Streptococcus agalactiae*: Implications for the microbial 'pan-genome'. Proc. Natl. Acad. Sci. U.S.A. *102*, 13950–13955.

Thomas, C.M., and Nielsen, K.M. (2005). Mechanisms of, and barriers to, horizontal gene transfer between bacteria. Nat. Rev. Micro. *3*, 711–721.

Tyson, G.W., Chapman, J., Hugenholtz, P., Allen, E.E., Ram, R.J., Richardson, P.M., Solovyev, V.V., Rubin, E.M., Rokhsar, D.S., and Banfield, J.F. (2004). Community structure and metabolism through reconstruction of microbial genomes from the environment. Nature *428*, 37–43.

Venter, J.C., Remington, K., Heidelberg, J.F., Halpern, A.L., Rusch, D., Eisen, J.A., Wu, D., Paulsen, I., Nelson, K.E., Nelson, W., *et al.* (2004). Environmental genome shotgun sequencing of the Sargasso Sea. Science *304*, 66–74.

Waldron, D.E., and Lindsay, J.A. (2006). Sau1: a novel lineage-specific type I restriction-modification system that blocks horizontal gene transfer into *Staphylococcus aureus* and between *S. aureus* isolates of different lineages. J. Bacteriol. *188*, 5578–5585.

Wirth, T., Falush, D., Lan, R., Colles, F., Mensa, P., Wieler, L.H., Karch, H., Reeves, P.R., Maiden, M.C.J., Ochman, H., *et al.* (2006). Sex and virulence in *Escherichia coli*: an evolutionary perspective. Mol. Microbiol. *60*, 1136–1151.

Zhaxybayeva, O., Gogarten, J.P., Charlebois, R.L., Doolittle, W.F., and Papke, R.T. (2006). Phylogenetic analyses of cyanobacterial genomes: Quantification of horizontal gene transfer events. Genome Res. *16*, 1099–1108.

The Extent and Regulation of Lateral Gene Transfer in Natural Microbial Ecosystems

6

Rustam I. Aminov

Abstract

The importance of horizontal gene transfer (HGT) in bacterial evolution is evident from the retrospective analyses of bacterial genomes, which suggest that a substantial part of bacterial genomes is of foreign origin. Another line of evidence that supports the possibility of rapid adaptation of bacteria through lateral gene exchange is the history of antibiotic use by humans. Within a very brief period of the 'antibiotic era' many bacterial pathogens were able to acquire the mechanisms allowing them to withstand the selective pressure of antibiotics. And, finally, field and microcosm studies allowed monitoring HGT events *in situ*. In this chapter, a brief overview of the milestones of mobile genetic elements (MGEs) research is given, followed by discussion of the conceptual framework development. Then the occurrence and diversity of MGEs as well as the frequencies of HGT in terrestrial, aquatic, intestinal and biofilm communities are described. The role of environmental factors that may affect MGE-mediated HGT in these ecosystems is also discussed.

Introduction

The first horizontal gene transfer experiments were performed by Griffith (1928), who 'transformed' rough, avirulent live pneumococci into smooth, virulent pneumococci by adding factors from dead, smooth and virulent pneumococci. The word 'transformation' at his time had no genetic connotation in the sense we would perceive it today; it was a 'transformation' of the pneumococcal phenotype. The nature of the transforming factor remained unknown until the 1940s, when it was recognized as DNA (Avery *et al.*, 1944) and, since then, the term 'transformation' acquired a genetic context. The subsequent discovery of bacteriophages (reviewed in Duckworth, 1976) and plasmids (Lederberg, 1952) has laid the foundation for the development of contemporary molecular biology, which has contributed enormously to many fields of biology, medicine, and biotechnology. The understanding of the fundamentals of the structure-and-function of mobile genetic elements as well as the enzymes of DNA metabolism allowed humans to adapt these mechanisms to serve their own needs. In the form of genetic engineering technology, these mechanisms are actively employed in a number of applications in science, biotechnology, and everyday life.

In a parallel development, the discovery and application of antibiotics beginning from the early 1940s has revolutionized the branch of clinical microbiology aimed at control and treatment of infectious diseases. Indeed, antibiotics are one of the most successful forms of chemotherapy known in medicine. Infectious diseases, which were the leading causes of human morbidity and mortality for most of the human existence, were placed under control, and many millions of lives were saved thanks to antibiotic interventions. Initially rare, antibiotic resistance among pathogens emerged as a major problem in the 1950s when Japanese hospitals encountered *Shigella* dysentery outbreaks that resisted to treatment by usual antibiotics. Intensive clinical and genetic investigations performed by Japanese scientists during that period resulted in a concept of episome-mediated transfer of drug resistance in the Enterobacteriaceae (Watanabe and Fukasawa, 1961; Watanabe, 1963). Moreover, this research also resulted in the eventual use of antibiotic resistance markers for genetic manipulations and was, therefore, one of the cornerstones for the development of genetic engineering.

Clinical studies focused on molecular mechanisms of antibiotic resistance, genetic elements involved and epidemiology have been developing in parallel to, but not interactively with, the environmental studies of HGT. The first attempts to estimate the frequencies of HGT in natural settings were performed in the 1970s with the use of model bacteria, *E. coli* and *Bacillus subtilis* (Weinberg and Stotzky, 1972; Graham and Istock, 1978). Since then, the approach involving the microcosm-based and field studies as well as a range of MGEs has resulted in a better understanding of how the environmental factors contribute to HGT in natural ecosystems (Van Elsas *et al.*, 2000; Timms-Wilson *et al.*, 2001). The analysis of gene exchange processes in natural settings was facilitated by the introduction of a variety of molecular ecology tools into the microcosm and field studies. These were the markers that allowed distinguishing the donor, recipient and transformant/transconjugant/transductant populations; DNA hybridization and sequencing; PCR typing; and others (Akkermans *et al.*, 1995; Götz *et al.*, 1996; Smalla *et al.*, 2000; Timms-Wilson *et al.*, 2001). The visualization of gene transfer *in situ* was made possible with the use of the *gfp* gene (Christensen *et al.*, 1996; Dahlberg el., 1998a; Andersen *et al.*, 1998). These technological advances allowed estimating the actual rates of HGT and factors affecting these rates in natural ecosystems.

Further substantiation for the important role played by HGT in the evolutionary process, especially among the Bacteria, came during the last decade, which is commonly referred to as the 'omics' era. With the advent of high-throughput sequencing it becomes possible to determine the genomic structure of many living organisms and apply the retrospective approach to HGT studies. Comparative genomic analyses revealed that, besides the core genes encoding essential cellular functions, the substantial part of bacterial genomes consists of auxiliary genes acquired by horizontal gene transfer (Ochman *et al.*, 2000). The latter group of genes may confer adaptive advantages under certain growth or environmental conditions that may contain antimicrobials, xenobiotics, heavy metals, sucrose, and other compounds. These genes also confer the important characteristics allowing the colonization of new ecological niches governed by biotic factors such as symbiotic and pathogenic relationships (Koonin *et al.*, 2001).

Conceptual and classification developments

The range of MGEs involved in the evolution and adaptation of bacteria through HGT is continuously updated and re-classified according to a better understanding of HGT mechanisms. In addition to the 'classical' MGEs such as plasmids, bacteriophages and transposons, the list now includes various genomic islands conferring pathogenic and symbiotic properties, large megaplasmids comparable in terms of size with bacterial chromosomes, and other elements with less defined structure-and-function. Owing to the ever extending list of newly described MGEs, it has been proposed to unify the heterogeneous classes of MGEs, such as conjugative transposons, integrative 'plasmids', genomic islands and numerous unclassified elements into integrative and conjugative elements (ICEs) (Burrus *et al.*, 2002). The basis for this re-classification was that these elements share similar characteristics, such as the excision by site-specific recombination, transfer by conjugation and integration by recombination between a specific site of elements and a site in the host's genome. Thus, ICEs combine the features of other MGEs, such as bacteriophages (integration into and excision from the host chromosome but no transmission by conjugation), insertion sequences (IS) and simple transposons (integration into and excision from the chromosome but no horizontal transfer), and conjugative and mobilizable plasmids (transferred from cell to cell by conjugation but replicated autonomously) (Wozniak and Waldor, 2010). A recent, even more radical, approach is to bring a number of MGEs under the umbrella of genomic islands (GEIs) (Juhas *et al.*, 2009). According to this view, GEI would encompass other categories of elements, such as ICEs, integrated plasmids, non-replicative but excisable elements (non-replicating *Bacteroides* unit, NBU; Shoemaker *et al.*, 2000), and even cryptic or damaged prophages.

Another interesting class of MGEs is the so-called gene transfer agent (GTA), which was initially described in *Rhodobacter capsulatus* (formerly *Rhodopseudomonas capsulata*) (Yen *et al.*, 1979) but later was found in many other bacteria (Lang and Beatty, 2007; Stanton, 2007). Although GTAs are similar to bacteriophages, they differ from them in two important aspects: (i) GTAs function only in genomic DNA transfer between cells; and (ii) there are no negative effects associated with gene transfer to the recipient. The representatives of another class of MGEs, named IS*CRs*, are known to move by a process called 'rolling-circle replication', and a function of this process is the concomitant movement of adjacent DNA upstream of their transposase genes (Toleman *et al.*, 2006). A recently discovered novel mechanism of HGT in *E. coli*, which was named 'cell-to-cell transformation', involves cell-derived DNA and, similar to Gram-positive bacteria, a putative promoting pheromone (Etchuuya *et al.*, 2011). These examples suggest that there is a continuum of MGEs, partially overlapping in their structure-and-function, rather than separate and strictly defined MGE classes.

Even the concepts within the well-established areas of MGE research such as plasmid biology may require further clarifications and adjustments in the light of novel information coming from a broader sampling of microbial diversity and the corresponding genetic mechanisms associated with this diversity. For example, the original classification of plasmids into compatibility groups was based on the ability of plasmids to stably coexist in the same bacterium (Datta and Hedges, 1971). Molecular mechanisms underlying the incompatibility were explained essentially as a stochastic phenomenon based on the inability of a bacterial host to differentiate among similar replicons and thus giving rise to homoplasmid segregants as a direct consequence of random assortment during cell division (Novick, 1987). Within the frames of this concept, it is hard to explain the stable inheritance of multiple large plasmids in the Alphaproteobacteria (Pradella *et al.*, 2010). The implementation of a phylogenomic approach to plasmid classification suggests that the Rhodobacterales harbour a set of at least 18 compatible plasmids, which can in principle stably coexist within the same cell (Petersen *et al.*, 2009, 2011). Another concept that is presently undergoing an extensive reassessment is the role of prophages in bacterial genomes. The conventional view of prophages as time bombs that are ready to enter the lytic cycle and destroy the host is gradually shifting towards their perception as a key component in bacterial survival strategies (Paul, 2008). Indeed, the prophages may confer many beneficial properties to the host helping to withstand osmotic, oxidative and acid stresses, as well as contributing to increased growth and biofilm formation (Wang *et al.*, 2010). In mammals, herpesvirus latency confers a symbiotic protection against infection with the bacterial pathogens *Listeria monocytogenes* and *Yersinia pestis* (Barton *et al.*, 2007).

The recent record of rapid MGE-mediated dissemination of antibiotic resistance genes among bacterial pathogens also gave a substantial boost to revisit the conceptual framework of antibiotics and antibiotic resistance. Historically, this research was an enclave within the boundaries of clinical microbiology and bacterial genetics because it was thought that the antibiotic resistance problem was due to, and could be resolved within, the antibiotic treatment practices of human infectious diseases. Also, a contributing factor to the isolation was the little interaction with other fields of microbiology. The realization of the point that antibiotic resistance research should take a broader than the clinical microbiology approach to identify the factors ultimately leading to the acquisition of antibiotic resistance by pathogens entered the mainstream thinking substantially later. In particular, the placement of the problem within the evolutionary and ecological contexts appeared to be especially fruitful (Davies *et al.*, 2006; Aminov and Mackie, 2007; Fajardo and Martínez, 2008; Aminov, 2009, 2010). Two main arguments emerged from this synthesis: firstly, it is the enormous diversity of antibiotic resistance genes existing in the environmental microbiota that has accumulated during billions of years of evolution; and, secondly, the realization that there are no barriers among the ecological compartments in the microbial world, and that the microbiota of different compartments may easily exchange the gene pool through MGE-mediated HGT.

Potential obstacles in HGT research

While exploring the wealth of data collected by environmental genomics/metagenomics we may get a glimpse of the HGT events of the past, we also understandably wish to know the extent of HGT in the present-day ecosystems, especially in the face of pressing needs such as the acquisition of antibiotic resistance by pathogens or the potential risks associated with genetically modified organisms (GMOs). Many microcosm and field experiments that are aimed at the estimation of HGT rates in natural ecosystems rely on model organisms with well-developed genetics. These organisms, however, have been selected as models owing to their features that are important for working under laboratory conditions, such as the simplicity of cultivation, tolerance to high nutrient concentration, the ease of genetic manipulation, and other characteristics that are not necessarily significant for the survival and multiplication in natural ecosystems. Laboratory strains of *Bacillus subtilis,* for example, have been selected and manipulated to become highly competent, which is important for bacterial genetics studies (Anagnostopoulos and Spizizen, 1961). Many wild-type strains of *B. subtilis,* however, are not naturally competent, and the transformation of these 'undomesticated' strains is possible only by highly unnatural means such as protoplast electroporation (Romero *et al.*, 2006).

Another problem which may be encountered when accessing HGT in natural ecosystems is that different MGEs may potentially interact with each other, and thus it is important to know the diversity of indigenous MGEs in microcosm and field studies. And finally, owing to dispersion and population sizes, microorganisms can be transferred across distant environments (Hooper *et al.*, 2008). This geographical component may substantially contribute to HGT, especially if the genetic material transferred confers the traits important for the survival and reproduction in the 'donor' ecological niche. Thus, a certain degree of caution must be exercised when estimating the frequencies of HGT in natural ecosystems.

In the following two parts of this chapter, HGT is viewed mostly from an ecological and environmental regulation perspective rather than focusing on genetics and regulation at the cellular level. In the first part, the soil, aquatic, gut and biofilm communities are discussed in respect to HGT processes in these ecosystems. In the second part, factors potentially affecting HGT in natural ecosystems are analysed.

HGT in natural ecosystems

Soil microbiota

Soil ecosystems harbour an extremely broad diversity of microbiota reflecting plant type, soil type, soil management regime, and a number of other biotic and abiotic factors (Garbeva *et al.*, 2004). To begin with, the physical environment of soil is very heterogeneous in terms of gaseous, liquid or solid phases (Smiles, 1988). Additional variables include abiotic factors such as temperature, pH, concentration of nutrients and oxygen and moisture content and biotic factors such as antagonistic, commensal, mutualistic and other relationships among the soil inhabitants. Since the regulation of genetic machinery is highly responsive to the environment, these factors may indeed affect the frequencies of HGT. For example, the rate of conjugal plasmid transfer in soil varies depending on abiotic factors, such as soil moisture and temperature (Richaume *et al.*, 1989), pH (Rochelle *et al.*, 1989), and soil type (Richaume *et al.*, 1992). Water movement in soil may influence the survival and transport of genetically engineered strains (Trevors *et al.*, 1990). As for biotic factors, the presence of earthworms, protozoa and fungi certainly affects the conjugal plasmid transfer in soil (Daane *et al.*, 1996, 1997; Sørensen *et al.*, 1999, 2000). In general, bulk soil is generally poor in nutrients, and the active metabolic state of bacteria required for HGT is possible in nutritional hot spots such as the rhizosphere, the phyllosphere, decaying plant and animal tissue, and in manure-applied soil (Van Elsas and Bailey, 2002).

Indeed, conjugative plasmid transfers among plant phytosphere inhabitants have been detected on multiple occasions (Van Elsas *et al.*, 1988; Lilley *et al.*, 1994; Björklöf *et al.*, 1995; Pukall *et al.*, 1996; Lilley and Bailey, 1997; Kroer *et al.*, 1998). The most important factors contributing to the frequency of horizontal plasmid transfer in the rhizosphere appeared to be root growth and exudate production (Mølbak *et al.*, 2007). Transfer ratios were about 10 times lower in control soil than in the pea and barley rhizospheres. For some plasmids, an exceptionally broad range of recipients in the rhizosphere was demonstrated (Musovic *et al.*, 2006). A high-frequency conjugal transfer of an IncP-1 plasmid pKJK10 was shown not only to bacteria belonging to the alpha, beta, and gamma subclasses of the Proteobacteria, but also to *Arthrobacter* sp., a Gram-positive member of the Actinobacteria. Compared with filter mating, the plant model (alfalfa sprouts) provided an environment substantially enhancing the transfer of plasmid- and transposon-encoded antibiotic resistance markers between lactic acid bacteria (Toomey *et al.*, 2009).

Application of manure to soil is another hot spot contributing to the increase in local concentrations of MGEs and the frequency of HGT in this ecosystem. This practice, as it has been shown, may enhance plasmid mobilization and survival of the introduced donor strain (Götz and Smalla, 1997). If applied together with an antibiotic, the effect is synergistic and affects the transfer frequencies and composition of MGEs introduced with manure (Heuer and Smalla, 2007). Field application of piggery manure, which harbours a substantial reservoir of broad-host range plasmids conferring multiple antibiotic resistance genes, results in the dissemination of IncN, IncW, IncP-1, and pHHV216-like plasmids into agricultural soils (Binh *et al.*, 2008). Clinically relevant class 1 integrons are also introduced into soil via similar practices (Binh *et al.*, 2009).

Besides contributing to the accumulation of MGEs and to HGT increase, the application of manure to soil has other important consequences such as the generation of novel MGE diversity and dissemination of novel phenotypes (for example, resistance to recently introduced antibiotics) among bacterial populations. An example of this kind was described recently (Heuer *et al.*, 2009) and deserves a more detailed discussion. In this work, a novel plasmid type with 36% G+C content was captured from manure-treated soil microbiota by conjugation to *E. coli* recipient. The core of plasmids of this type is probably the product of recombination, comprising transfer and maintenance genes with moderate homology to plasmid pIPO2 and a replication module (*rep* and *oriV*) of other descent, correspondingly. These plasmids encode a number of antibiotic resistance genes including *tet*(X) (GenBank accession number FJ012881), which may confer resistance against a recently introduced third-generation tetracycline, tigecycline (Moore *et al.*, 2005). The species of *Acinetobacter* are the putative hosts for these low G+C plasmids in soil ecosystem (Heuer *et al.*, 2009), which is of concern because the closely related multidrug-resistant *Acinetobacter baumannii* is one of the currently emerging threats in hospitals (Dijkshoorn *et al.*, 2007). Tigecycline is currently proposed as a new treatment choice against *A. baumannii* (Bosó-Ribelles *et al.*, 2008) but this bacterium already poses a significant problem with the easily emerging resistance during tigecycline therapy, albeit conferred by a mechanism other than TetX (Peleg *et al.*, 2007; Damier-Piolle *et al.*, 2008). The similarity of structural and replication features of low G+C plasmids from soil to the corresponding plasmids of clinical *A. baumannii* isolates, together with the stable inheritance of these plasmids in *Acinetobacter* spp. (Heuer *et al.*, 2009), suggests that, once acquired, the plasmids may be easily accommodated by *A. baumannii* in the clinical environment. Given the carriage of resistance against the drug of last resort by these plasmids, their possible entry into clinical *A. baumannii* is a highly undesirable scenario. Disappointingly, there is some evidence that at least some clinically relevant resistance genes have an environmental origin (Wright, 2010) suggesting past horizontal gene exchange events between these ecological compartments. Thus, the microbiota of soil, especially of manure-fertilized soil, harbours a wide variety of MGEs enabling extensive HGT within and among microbial ecosystems.

At the same time, nutritionally poor soil ecosystems may also be involved in the HGT process. For example, a deep terrestrial subsurface, which is highly oligotrophic and extreme in terms of physical and chemical conditions, still harbours bacterial populations that carry MGEs and the genetic signatures of past gene transfer events, some of them apparently recent (Coombs and Barkay, 2004; Coombs, 2009). It is still more likely that the HGT rates under the nutritionally favourable conditions are higher owing to the substantial biosynthetic and energetic requirements for conjugation, DNA uptake, and lytic cycle.

Recent assessments of several types of soil revealed that lysogeny is relatively common in soil microbiota (Williamson *et al.*, 2007; Ghosh *et al.*, 2008; Williamson *et al.*, 2008). Similar to plasmid-mediated HGT, the diversity and dynamic of bacteriophages is also mostly confined to the nutritionally rich hot spots such as the rhizosphere. The decline in the introduced fluorescent *Pseudomonas* sp. population revealed the presence of large numbers of bacteriophage in the sugar beet rhizosphere (Stephens *et al.*, 1987). On the other hand, the growth and interaction dynamics of streptomycetes and a bacteriophage, which was investigated under a less nutritionally rich condition of soil microcosm, did not reveal such an effect of phages (Burroughs *et al.*, 2000). No measurable impact on the host in terms of reduced growth by the phage was found under these conditions. Interestingly, the burst size of the phage was larger in soil relative to that observed in liquid culture, suggesting that *in vitro* transduction experiments may underestimate the impact of this particular HGT mechanism in natural ecosystems.

In the soil ecosystem, bacteriophages display local adaptation to their bacterial hosts (Vos *et al.*, 2009). Sympatric phages are more infective than are phages from samples some distance away, suggesting that the phage-mediated HGT is highly localized in soil compartments. In model microcosm experiments, the presence of phages greatly reduced the sympatric diversity of the host bacterium but favoured the allopatric host diversification (Buckling and Rainey, 2002).

The possibility of natural genetic transformation of *B. subtilis* and *Pseudomonas stutzeri* has been demonstrated in a soil/sediment model system (Lorenz *et al.*, 1988; Lorenz and Wackernagel, 1990). These studies found significantly higher rates of transformation if the transforming DNA was associated with mineral/particulate material and thus protected against nuclease degradation. The protective effect of DNA absorption on minerals comprising soil and sediments against DNases has been noticed on many occasions (Aardema *et al.*, 1983; Romanowski *et al.*, 1991; Demanèche *et al.*, 2001; Cai *et al.*, 2006). While some naturally competent bacteria, as in the examples above, indeed demonstrate a certain level of transformability, the occurrence of naturally transformable bacteria amongst bulk and rhizosphere soil microbiota is very low (Richter and Smalla, 2007). Various transformation assays were performed, but only transformants with a positive control, *Acinetobacter baylyi* BD413, which is a highly transformable strain and can be transformed by DNA from virtually any sources, were obtained. Thus, the authors concluded that the proportion of native rhizosphere and bulk soil bacteria, which are naturally transformable, is negligibly low.

There have been (and indeed there still are) intensive debates surrounding the issue of possible risks associated with the release of genetically modified organisms (GMOs) into the environment, in particular, transgenic plants. Studies on persistence of plant DNA in soil arrived at different conclusions depending on the experimental setup. In microcosm experiments modelling *in situ* transport of recombinant plant DNA (rDNA) from Roundup Ready (RR) maize and soybean by leachate water, half-lives of rDNA in leachate water ranged from 1.2 to 26.7 h, depending on the temperature (Gulden *et al.*, 2005). The presence of transgenic DNA in soil where RR maize and soybean were cultivated can be detected by real-time PCR for up to 1 year after seeding (Lerat *et al.*, 2007). In soil microcosm experiments, recombinant DNA entry into soil from the decomposing RR leaf biomass was investigated, and the corresponding DNA was detectable in soil after 30 days (Levy-Booth *et al.*, 2008). At the present stage of our knowledge, HGT from transgenic plants to terrestrial bacteria is considered to be a rare event (Nielsen *et al.*, 1998). A recent study with the use of an *in situ* visualization technology, nevertheless, demonstrated this process could be observed in real time (Pontiroli *et al.*, 2009). This study, however, employed a highly transformable *A. baylyi* strain BD413, which may lead to the overestimation of natural transformation rates. The current view is that even if HGT from transgenic plants to soil microbiota takes place, it is not expected to influence prokaryotic evolution or have negative effects on human or animal health and the environment (Brigulla and Wackernagel, 2010).

Microbiota of aquatic ecosystems

More than 70% of the earth surface is covered with water, and the World Ocean is one of the principal components forming the climate and biosphere of the Earth. As the largest habit on the Earth, the World Ocean hosts a large diversity of life and is responsible for nearly the half of oxygen production. Besides, there are many freshwater ecosystems, such as lakes, rivers, and smaller water bodies.

One of the earliest accounts on the occurrence of antibiotic resistance in marine bacteria has found that this phenotype is common within this microbiota (Sizemore and Colwell, 1977). Interestingly, the resistance phenotype was more frequently encountered in bacteria from seawater samples collected offshore than for those collected near shore. Antibiotic resistance was even present in bacteria collected about 522 km offshore and at depths of 8200 m. In 6 out of 10 bacterial isolates the presence of plasmid DNA was confirmed by ethidium bromide-caesium chloride density gradient centrifugation. In general, however, antibiotic resistance phenotype and the presence of plasmids in marine bacteria tend to correlate with the degree of pollution, especially by toxic chemical waste (Baya *et al.*, 1986; Young, 1993). A recently applied cultivation-free approach for characterization of coastal *Synechococcus* metagenome relied on flow cytometry-based sorting of cells with a subsequent 454-shotgun pyrosequencing (Palenik *et al.*, 2009). This approach allowed identifying novel plasmids that were not found in model strain genomes of this clade. Plasmids with an enormous size range are also widespread in the *Roseobacter* clade, the representatives of which constitute up to 25% of the total marine bacterial community and thus play a global role in this ecosystem (Brinkhoff *et al.*, 2008). Some representatives may carry up to 12 extrachromosomal replicons, suggesting a very sophisticated partition mechanism ensuring their stable inheritance (Pradella *et al.*, 2010). In marine *Vibrio* the majority of plasmids are associated with pathogenic or symbiotic relationships with the host organism (Hazen *et al.*, 2010).

With the advent of high-throughput sequencing technologies, it becomes possible to characterize the marine phage diversity without the isolation and cultivation steps, which are extremely time- and resource-consuming while the outcome is limited to a few phages that we are able to multiply and characterize under laboratory conditions. In fact, the metagenomic approach is the only way to characterize the viral diversity because there is no single gene common to all phages to serve as a phylogenetic marker (Edwards and Rohwer, 2005). Comparative marine virome analyses involving four oceanic regions demonstrated that the global diversity is very high – presumably several hundred thousands of species (Angly *et al.*, 2006). The oceanic regions had different assemblages of marine phages suggesting strong local selective pressures enriching for certain viral types. Metagenomic characterization of viruses within aquatic microbial samples revealed a prevalence of genes encoding microbial physiological functions among viral sequences (Williamson *et al.*, 2008). Screening of 113 marine bacterial genomes for prophages yielded 64 prophage-like elements, 21 of which strongly resembled GTAs (Paul, 2008). Hence, viral- and GTA-mediated HGT is a common mechanism for generating microbial diversity in the marine environment thus contributing to survival in different parts of this extensive ecological niche. Moreover, marine phages may directly contribute to the fitness of their hosts. It has been suggested, for example, that the auxiliary genes in marine phages may confer selective advantage to the host through the viral gene cassettes encoding core photosystems, I and II (Sharon *et al.*, 2009). In a more recent work from the same group the repertoire of host genes detected in marine viruses appeared to be even broader including genes of energy metabolism as well as involved in translation and post-translational modification (Sharon *et al.*, 2011).

Historically, the *in situ* HGT rates in aquatic ecosystems have been typically studied with the use of microcosms, and it is understandable that in a number of cases these models have certain limitations in reproducing the vast range of ecological niches present, for example, in the World Ocean. In most cases, the microcosms are limited to modelling the shallow costal or estuarine ecosystems, frequently without the presence of ambient macro- and microbiota. Laboratory experiments also lack the scale of natural ecosystems and, as a consequence, may discount the role of important variables potentially affecting HGT. Another aspect, which needs to be taken into consideration, is the use of model donors and recipients that have been 'preselected' under laboratory conditions to monitor HGT. The HGT results obtained with the use of model organisms, genetic constructs and microcosms should be interpreted with a reasonable degree of precaution when extrapolating them in an attempt to describe a broader environment.

It has been shown that the presence of marine sediments in microcosms facilitates the uptake and expression of exogenous DNA by transformable marine *Vibrio* sp. (Stewart and Sinigalliano, 1990). It was concluded from this study that sediments were more likely niches for natural transformation than the water column in the marine environment. However, another study of natural plasmid transformation of a *Vibrio* strain in marine water column and sediment microcosms, in the presence of the ambient microbial community, arrived at the opposite conclusion (Paul *et al.*, 1991). The authors explained this discrepancy by a differential experimental setup involving the presence of ambient microbiota in their experiments, while the previous work used sterile sediments. Estimation of transformation rates in estuarine environments based upon the distribution of competency and transformation frequencies in isolates and mixed populations ranged from 5×10^{-4} to 1.5 transformants per day (Frischer *et al.*, 1994). Another aspect of natural transformation is that the structure-and-function of transforming DNA may be affected by the *in situ* microbiota. For instance, a broad-host-range plasmid pQSR50 that was introduced into the indigenous marine bacteria by natural transformation was subjected to the alteration of restriction profiles (Williams *et al.*, 1997). This involved changes in methylation patterns as well as structural rearrangements of the plasmid following gene transfer, thus contributing to the generation of plasmid diversity among the *in situ* bacterial populations.

Conjugative plasmid transfers under simulated marine environment conditions have been demonstrated in many experiments. Transconjugants can be detected even under oligotrophic conditions and at very low population densities of donors and recipients (Goodman *et al.*, 1993). The heterogeneity of the marine environment affects HGT rates; the plasmid transfer frequency is higher among cells attached to the bead surfaces in the biofilm than among cells in the aqueous phase (Angles *et al.*, 1993). The use of the *in situ* technology with the *gfp*-tagged conjugative plasmid pBF1 from *Pseudomonas putida* suggests that the directly determined rates of horizontal plasmid transfer in marine bacterial communities may be high, ranging from 2.3×10^{-6} to 2.2×10^{-4} transconjugants per recipient (Dahlberg *et al.*, 1998b).

In one of the initial studies, transduction of *P. aeruginosa* streptomycin resistance by a generalized transducing phage, F116, was monitored in a flow-through chamber suspended in a freshwater reservoir (Morrison *et al.*, 1978). The frequency of F116-mediated transduction ranged from 1.4×10^{-5} to 8.3×10^{-2} transductants per recipient during the 10-day incubation period. A recent investigation of phage-mediated gene transfer in freshwater environments used a more advanced technological tool to monitor these events at the single-cell level; the so-called Cycling Primed *in situ* amplification-Fluorescent *in situ* Hybridization (CPRINS-FISH) (Kenzaka *et al.*, 2010). The P1, T4 and EC10 phages mediated gene transfer from *E. coli* to both plaque-forming and non-plaque-forming Enterobacteriaceae strains at frequencies of $0.3–8 \times 10^{-3}$ per plaque-forming unit (PFU). The rate of transfer of EC10 ranged from undetectable to 2×10^{-3} per total direct cell count when natural bacterial communities were the recipients. This suggests that transduction may involve a wide range of bacteria, not necessarily limited by close relatives, and that the phage-mediated HGT is a frequent event in freshwater environments.

Compared to the freshwater microcosm studies, the corresponding experiments imitating the marine environment yielded lower frequencies of plasmid transduction in the mixed bacterial communities; the values were in the range from 1.58×10^{-8} to 3.7×10^{-8} transductants/PFU (Jiang and Paul, 1998). This is not to say that the overall HGT rates in marine ecosystems are low. Other MGEs residing in marine microbiota may generate an extraordinarily extensive gene flow – one of the highest ever detected. A recent study demonstrated extremely high rates of HGT in marine ecosystems that are mediated by GTAs (McDaniel *et al.*, 2010). The frequencies detected are by many orders of magnitude higher than those of transformation or transduction, with as many as 47% of the culturable marine microbiota being confirmed as gene recipients. Given the wide presence of GTAs in phylogenetically and ecologically diverse bacteria (Stanton, 2007), GTA-mediated HGT may be more common in other natural ecosystems and may generate more extensive gene exchange than previously anticipated.

Comparative genomics and transcriptomics of marine bacteria are consistent with the high rates of HGT in these ecosystems. The co-occurring *Shewanella baltica* isolates from similar depths exchanged a larger fraction of their core and auxiliary genome compared with the isolates from more different depths (Caro-Quintero *et al.*, 2011). These HGT events took place in the very recent past reflecting the rapid adaptation to environmental settings through gene acquisition. Genomic evolution for a cold marine lifestyle and *in situ* explosive biodegradation in *Shewanella* spp. also involved an extensive gene acquisition from deep-sea bacteria (Zhao *et al.*, 2010). A recent study of global population structure of *Vibrio cholerae*, which is a widely distributed microbial species in the coastal ocean regions, revealed a rapid diversification among geographically distant populations due to the extensive local gene exchange (Boucher *et al.*, 2011). This integron-mediated HGT involves mostly the gene pool encoding secondary metabolism and cell surface modification thus allowing rapid adaptation to local ecological niches.

Most of our knowledge about the HGT events in aquatic ecosystems comes from studies of the resident bacteria and their corresponding MGEs. Much less is known, however, about the corresponding mechanisms and events taking place in the other procaryotic domain of life, Archaea. The explanations for this may be due to a much smaller absolute number of representatives from this taxonomic group as well as their adaptation to extreme environments making it difficult to study them under laboratory conditions. Nevertheless, the advent of cultivation-free approaches such as metagenomic analysis allowed uncovering the past HGT events engraved in the archaeal genomes. Detailed phylogenetic analysis of the 41 genes present in a genome fragment of a marine planktonic group I archaeon, DeepAnt-EC39, revealed that 11 of them had been most likely acquired by HGT from different sources, including bacteria and other unrelated archaea (López-García *et al.*, 2004). The possession of a proteorhodopsin gene by marine planktonic archaea is also likely owing to the past acquisition from bacteria (Frigaard *et al.*, 2006). Metagenomic analysis of marine uncultured planktonic archaea revealed frequent HGT acquisition, mostly from bacterial donors but also from other archaea and eukaryotes (Brochier-Armanet *et al.*, 2011). Interestingly, the laterally transferred gene pool is conserved within the several representatives of the studied groups suggesting the important role played by HGT in the adaptation of these archaea to the cold and nutrient-depleted deep marine environment. Genomic plasticity of the psychrophilic archaeon, *Methanococcoides burtonii*, driven, in particular, by HGT and transposase activity is also thought enabled its adaptation to the cold Antarctic lake environment (Allen *et al.*, 2009). On the other side of the temperature extreme, transposition, genome rearrangements and HGT events are also seen as a necessary adaptation of *Pyrococcus* species to the shallow hydrothermal vent ecosystem (Hamilton-Brehm *et al.*, 2005; White *et al.*, 2008). In general, archaeal populations of aquatic ecosystems demonstrate the same tendency as bacteria to form local mobilomes and gene pools, allowing a better adaptation to a particular ecological niche.

Intestinal microbiota

The frequencies of horizontal gene transfer by conjugation are usually estimated using standard mating techniques, in liquid culture or on the surface of solidified media or filters. The microcosm and field experiments that are the better estimates of HGT in the environment usually deal with lower population densities, lower nutrient availability, and generally lower temperatures. The *in vivo* conditions, however, are substantially different, especially in the intestine, where the enormously dense and diverse microbiota performs a number of functions important for the host organism. These functions include the prevention of colonization by pathogens, degradation of dietary (polysaccharides) and *in situ*-produced (mucin) compounds, production of nutrients (short chain fatty acids and vitamins), shaping and maintenance of normal mucosal immunity, and contribution to intestinal epithelial homeostasis. The commensal microbiota is under constant surveillance by the innate (antimicrobial peptides) and adaptive (immunoglobulins) immunity. The effect of innate immunity breaches on gut microbiota can be seen in genetically susceptible hosts, the commensal microbiota of which is formidably restructured compared to normal subjects (Khachatryan *et al.*, 2008).

The discovery of ICEs among the representatives of the Bacteroidetes, the major bacterial phylum in the mammalian gut, brought the notion that gene transfer in this ecosystem may be intense (Salyers, 1993), and gut microbiota, therefore, may represent one of the major reservoirs for antibiotic resistance genes (Salyers *et al.*, 2004). Indeed, the taxonomically different representatives of gut microbiota may share the pool of closely related antimicrobial resistance genes (Frye *et al.*, 2011). The role of bacteriophages in acquisition of genes by pathogenic gut microbiota and subsequent evolution towards increased pathogenicity also pointed to HGT as a crucial event in the development of virulence traits and antibiotic resistance (Calderwood *et al.*, 1987; Barondess and Beckwith, 1990; Brabban *et al.*, 2005). More recent metagenomic studies of the human gut microbiota allowed estimating the role of MGEs without the cultivation bias. The uncultured viral community from human faeces contained an estimated 1200 viral genotypes (Breitbart *et al.*, 2003). Interestingly, unlike the similarity of gut bacterial communities among genetically related individuals, the viral communities are unique to individuals regardless of their degree of genetic relatedness (Reyes *et al.*, 2010). Besides viromes, a recent metagenomic inventory identified a conjugative transposon family explosively amplified in human gut microbiomes (Kurokawa *et al.*, 2007). The culture-independent TRACA (Transposon-Aided Capture) system was used to sample plasmid diversity in the human gut microbiota (Jones *et al.*, 2010). This study suggested a broad global distribution of some plasmids and plasmid families that are potentially unique to the human gut microbiome. Thus these findings further supported the earlier cultivation-based views that the intestinal microbial ecosystem is extremely enriched by MGEs thus making it the arena of a potentially extensive gene exchange. Indeed, the range of genes that have been exchanged in the past is not limited to the virulence and antibiotic resistance genes but also includes the genes of the core gut microbiome such as those encoding bile salt hydrolases (Elkins *et al.*, 2001) or butyrate metabolism enzymes (Louis *et al.*, 2007).

Humans and agricultural animals are the main consumers of antibiotics, for therapeutic, prophylactic and growth-promoting purposes and, as it will be discussed in the next part, antibiotics may substantially increase the rates of HGT. But what could be other host-mediated factors contributing to the maintenance and transfer of MGEs in gut microbiota? It seems that the *in vivo* environment itself may enhance the transfer frequency and contribute to the stable inheritance of MGEs even in the absence of selection by antibiotics (Johnsen *et al.*, 2002; Dahl *et al.*, 2007). Thus, *in vitro* models may substantially underestimate the transfer potential of MGEs. In nutritionally poor environments, such as bulk soil, the presence of earthworms greatly enhances the transfer of plasmid pJP4 from an inoculated donor bacterium, *P. fluorescens*, to the indigenous soil microbiota (Daane *et al.*, 1996). Other soil microcosm experiments, modelled with *E. coli* as a donor of a genetically marked large conjugative plasmid RP4luc, in the presence or absence of earthworms, provided evidence that the gut passage was a precondition for a plasmid transfer to soil microbiota (Thimm *et al.*, 2001). Interestingly, the plasmid was transferred at higher frequencies than detected in filter mating, suggesting once again that the HGT rates in nature may be higher than the laboratory estimates. This observation also confirms the earlier notion that microbial ecosystems are not isolated and there is a potential for lateral gene exchange among different microbial ecosystems. If MGEs from soil have entered the earthworm gut, then they can also enter the gut of animals that are next in the food chain, for example, moles and birds.

Another curious factor that may contribute to enhanced HGT is the presence of ciliates. Ciliates are common in many aquatic ecosystems as well as in the gastrointestinal systems such as the rumen. Their food vacuoles are formed through phagocytosis and follow a particular path through the cell resembling a primitive gastrointestinal tract. It has been shown that ciliates may enhance the rate of conjugal transfer between *E. coli* strains by two orders of magnitude, and the mechanism involved is the accumulation of bacteria in vesicles (Matsuo *et al.*, 2010). The mechanism described may contribute to the dissemination of antibiotic resistance in bacterial populations (Oguri *et al.*, 2011).

The insect gut can also be considered as a hot spot contributing to the enhanced HGT in this ecosystem. For example, the rates of conjugative plasmid transfer between *S. enterica* Newport and *E. coli* in the gut of the lesser mealworm beetle are by two orders of magnitude higher compared to filter mating (Poole and Crippen, 2009). The occurrence of conjugal plasmid transfer and transduction was also observed in the flea and house fly gut (Hinnebusch *et al.*, 2002; Petridis *et al.*, 2006; Akhtar *et al.*, 2009). An assessment of natural transformability of bacteria in the insect gut, however, failed to detect any transformation event, even with the use of *A. baylyi* strain BD413 as a recipient (Ray *et al.*, 2007), thus suggesting that natural transformation plays a minor role in HGT in the gut of insects and, possibly, in the gut of other animals.

Conditions in the gut can be considered as very favourable for HGT. Firstly, the host provides a continuous inflow of nutrients that allow maintaining the active metabolism of gut microbiota. Secondly, the population densities are extremely high and thus conducive to the HGT mechanisms requiring intimate cell-to-cell contact such as conjugation. Thirdly, in a part of animals, with homeothermic metabolism, the constant body temperature may allow the bacterial cells to perform at optimal efficiency. And finally, the vast diversity of gut microbiota itself may have an 'amplification effect' for HGT (Dionisio *et al.*, 2002). In addition to these well-known factors, recent investigations have uncovered the mechanisms of host-microbe molecular crosstalk that may contribute to the frequencies of HGT in the gut (Mulder *et al.*, 2011).

One of these mechanisms is based on the ability of bacteria to sense and respond to host signals. In particular, bacterial sensing and responding to the level of host stress hormones is a well-established fact (Sperandio *et al.*, 2003; Clarke *et al.*, 2006; Karavolos *et al.*, 2008; Spencer *et al.*, 2010). Bacterial responses to the host stress may also involve a genetic component, which is expressed through the enhanced conjugative gene transfer between enteric bacteria (Peterson *et al.*, 2011). In these *in vitro* experiments, the physiological concentrations of noradrenaline stimulated the transfer of a conjugative plasmid from a clinical strain of *Salmonella* sp. to an *E. coli* recipient. Interestingly, the adrenergic receptor antagonists negated the stimulatory effect of noradrenaline on conjugation. These mediators of host stress may possibly affect HGT under the *in vivo* environment as well.

The issues of potential risks associated with the consumption of GMOs by humans and animals have been addressed in a number of feeding trials. In general, there is a lack of evidence that DNA of transgenic plants, in particular the markers used for their construction, can be taken up by gut microbiota or enter the organs other than the gastrointestinal tract. Neither small fragments of transgenic DNA nor immunoreactive fragments of transgenic protein were detectable in loin muscle samples from pigs fed a diet containing Roundup Ready soybean meal (Jennings *et al.*, 2003). An assessment of the survival of transgenic plant DNA in the human gastrointestinal tract concluded that gene transfer did not occur during the feeding experiment involving GM soya (Netherwood *et al.*, 2004). No traces of the construct or endogenous soybean DNA could be detected in muscle samples of rats fed soybean meal from Roundup Ready or conventional soybeans (Zhu *et al.*, 2004). Likewise, no traces of transgenic DNA were detected in the milk of cows fed maize silage from an herbicide-tolerant genetically modified variety (Phipps *et al.*, 2005). Plasmid and genomic DNA from GM plants were used in *in vitro* and *in vivo* (mono-associated rats) transformation studies, but no detectable transfer of DNA was found (Wilcks and Jacobsen, 2010). Attempts to detect DNA transfer from transgenic plants to bacteria in the intestine of the tobacco hornworm (Deni *et al.*, 2005) or bees were also unsuccessful (Mohr and Tebbe, 2007).

Microbiota of biofilms

Biofilms are the matrix-enclosed aggregates of microbiota, attached to each other and to biological or non-biological surfaces (Hall-Stoodley *et al.*, 2004). This ancient form of adaptation appeared very early in the prokaryotic phase of evolution being selected as a trait helping to withstand the hostile environmental factors. Indeed, biofilms can be perceived as a form of primitive multicellular organisms, which use the same strategy in their interaction with the environment as their eukaryotic counterparts do. This successful trait, therefore, was replicated in many bacterial and archaeal lineages; and biofilm communities are widespread in many natural ecosystems. It is now widely acknowledged that the majority of microbiota found in natural, clinical, and industrial settings persist in association with surfaces rather than in the planktonic state (Costerton, 1995; Davey and O'Toole, 2000). In natural ecosystems, biofilms are usually found in many aquatic or semi-aquatic ecosystems such as rocks and pebbles of most streams and rivers, on the surface of still water bodies, in wastewater treatment systems, in water and sewage pipes, in hot springs, in the subtidal and intertidal solid surfaces of marine ecosystems, on the teeth of humans and animals, in chronic infections of human body, and in many other ecosystems. For studies of HGT in highly organized structures as biofilms, technologies such as single-cell detection of donor, recipient and transconjugant bacteria, combined with individual-based mathematical models, have been developed to estimate the HGT rates *in situ* (Sørensen *et al.*, 2005). Raman spectrometry and environmental scanning electron microscopy analyses, combined with molecular ecology tools, allow a better understanding of structure-and-function of biofilms including the chemical composition of the matrix, microbiota embedded in the matrix, and the spatial distribution of biofilms (Schwartz *et al.*, 2009).

Most of our knowledge in regards to the biology of biofilms came from clinical microbiology. This increased attention to the clinical aspects of biofilm communities was primarily dictated by their role in human disease (Parsek and Singh, 2003; Fux *et al.*, 2005; Lindsey and von Holy, 2006; Estrela *et al.*, 2009; Kaplan, 2010). Biofilm formation is linked to the pathology of many infectious diseases, and biofilms are notoriously difficult to eradicate because of the increased level of resistance to antibiotics (Stewart and Costerton, 2001; Anderson and O'Toole, 2008; Høiby *et al.*, 2010a). The ability to form highly resilient biofilms has been demonstrated for many human pathogens such as *Pseudomonas aeruginosa* (for recent reviews, see Harmsen *et al.*, 2010; Hassett *et al.*, 2010; Häussler, 2010; Høiby *et al.*, 2010b), *Staphylococcus aureus* (Goerke and Wolz, 2010), *S. epidermidis* (Fey and Olson, 2010; Rohde *et al.*, 2010), *Streptococcus pneumoniae* (Moscoso *et al.*, 2009), *S. mutans* (Senadheera and Cvitkovitch, 2008), *Neisseria gonorrhoeae* (Greiner *et al.*, 2005; Falsetta *et al.*, 2009; Steichen *et al.*, 2011), *Campylobacter jejuni* (Haddock *et al.*, 2010; Naito *et al.*, 2010), *Candida* spp. (Morales and Hogan, 2010; Williams *et al.*, 2011), and others. Besides, microorganisms may form mixed-species communities and, in a number of cases, this may promote the development of biofilms (Bamford *et al.*, 2009; Silverman *et al.*, 2010; Teh *et al.*, 2010).

Given their structure-and-function, biofilms are the hot spots for HGT because they provide high population densities and close proximity of cells, cells in biofilms are metabolically active, and the microbiota is protected against harsh environment, predators and immune surveillance by the extracellular matrix in which the cells are encapsulated. Indeed, HGT occurs with enhanced efficiency in biofilms, and conjugative plasmids themselves contribute to the development, stabilization and expansion of biofilms (Hausner and

Wuertz, 1999; Ghigo, 2001; Molin and Tolker-Nielsen, 2003; Reisner *et al.*, 2006; Burmølle *et al.*, 2008). Conjugal plasmid transfer that is implemented through the synthesis of pili and type IV secretion system contributes to the intimate cell-to-cell contact, thus facilitating the formation and growth of biofilms. Type IV secretion systems use a pilus-based system to mediate DNA or protein transfer (Hayes *et al.*, 2010). The involvement of type IV secretion system in biofilm formation has been demonstrated for a number of bacteria (Shime-Hattori *et al.*, 2006; Li *et al.*, 2007; Luke *et al.*, 2007; Barken *et al.*, 2008; Varga *et al.*, 2008; Bahar *et al.*, 2009; Gibiansky *et al.*, 2010). Protection under the umbrella of biofilms confers a selective advantage for the bacteria as well as aids further plasmid/ICE/GEI transfers. Thus, this positive feedback sustains the diversity of MGEs in biofilms and offsets the fitness cost associated with the carriage of MGEs.

The presence of prophages in the host's genome may have a modulatory effect on biofilm formation and physiology of the host, including central metabolism (Wang *et al.*, 2009). Some prophage-encoded proteins are actually essential for biofilm formation (Toba *et al.*, 2011). The modulatory effect of prophages is expressed in the enhancement of biofilm formation and the resistance of biofilms against adverse environmental conditions (Carrolo *et al.*, 2010; Wang *et al.*, 2010). The biofilm matrix is a conglomerate of polymers usually consisting of extracellular DNA (eDNA), proteins, and polysaccharides. In this regard, a partial phage-mediated lysis of a proportion of microbiota in biofilm, due to spontaneous phage induction, may provide the eDNA supply, thus contributing to biofilm formation and maintenance by the remaining bacterial populations (Carrolo *et al.*, 2010; Gödeke *et al.*, 2011). At the same time, transmission electron microscopy reveals also the presence of intact bacteriophage particles that are enmeshed in the extracellular polymeric matrix of biofilms (Kay *et al.*, 2011). Precise deletion of *E. coli* prophages uncovered the role of e14 and rac proteins in the increased early biofilm formation (Wang *et al.*, 2010). Besides, other prophages contributed to the host resistance against a variety of stresses, including osmotic, oxidative and acid. These examples reveal the role of the phage-mediated HGT in bacterial evolution and adaptation, especially in regards to the acquired ability to form biofilms as well as to withstand environmental stresses.

Natural competence and transformability in bacteria usually correlates with the presence of type IV pili or type IV pilin-like proteins (Averhoff and Friedrich, 2003). While the role of conjugation machinery and pili in biofilm formation and subsequent acceleration of conjugative element transfer is firmly established, it is not clear if the same is true for natural transformation. Given the involvement of type IV pili or type IV pilin-like proteins in natural transformation and in biofilm formation, would this result in a higher rate of transformation within biofilms? Do the large quantities of eDNA in the extracellular matrix participate in natural transformation of bacteria in biofilms? There are very few works that have addressed these questions. Horizontal transfer of non-conjugative plasmids in *E. coli* colony biofilms suggested the possibility of natural transformation within bacterial biofilms (Maeda *et al.*, 2006). Later, however, it became clear that the mechanism of DNA acquisition in the biofilm is more complex and was, therefore, described as a 'cell-to-cell transformation', which also involves a peptide pheromone as a regulator (Etchuuya *et al.*, 2011). Other circumstantial evidence of DNA transfer due to close cell-to-cell contact (Van Randen and Venema, 1984; Wang *et al.*, 2007) may be interpreted within the frames of this hypothesis, although the transfer in these cases was between the genera, and the role of pheromones was not shown. The language of intercellular communication used in regulation of various processes within the complex structure of biofilms is indeed the language of pheromones such as N-acyl-L-homoserine lactones, furanosyl borate diester and peptide autoinducers (Dickschat, 2010). Recently, the possibility of transfer of tetracycline resistance by transformation with eDNA within model oral biofilms was described, although without specifying the frequency of such an event (Hannan *et al.*, 2010).

Factors influencing HGT

Stress and SOS response

It has been long recognized that various stress conditions may contribute to increased rates of HGT. For example, UV irradiation or starvation affects the mobility of transposons and insertion sequences (Levy *et al.*, 1993; Ilves *et al.*, 2001). While the lysogenic cycle in phage development limits its inheritance to the vertical transfer within the chromosome of the host, the stress factors such as induction of the SOS response or amino acid deprivation of the host lead to the lytic cycle and further horizontal dissemination of phage and transduced DNA (Melechen and Go, 1980; Little, 2005). Control of horizontal transfer of ICEs, which are found in many bacterial genomes and which encode a variety of properties besides antibiotic resistance and virulence, may also be regulated through stress, SOS and other environmental signals (Auchtung *et al.*, 2005; Bose *et al.*, 2008). In particular, the SOS response, which is induced by DNA damaging agents such as mitomycin C and antibiotics such as fluoroquinolones and dihydrofolate reductase inhibitors, leads to the expression of the SXT activators resulting in more than 300-fold increase of HGT rates (Beaber *et al.*, 2004). Thus the consequences of using SOS response-inducing antibiotics may result in co-selection of other antibiotic resistance genes that are physically linked in a mobile genetic element (Hastings *et al.*, 2004). Moreover, the SOS response controls integron recombination thus enhancing the potential for cassette swapping and capture in cells (Guerin *et al.*, 2009). Besides the antibiotic- and SOS response-regulated excision, transfer and integration of ICEs, regulatory factors include the stationary phase of growth and the presence of 3-chlorobenzoate (Sentchilo *et al.*, 2003). Interestingly, this chlorinated compound stimulates horizontal transfer of the genes encoding its own metabolism. This may explain the previous results of the authors that suggested the need of specific substrates for genetic transfer to occur in activated-sludge microcosms (Ravatn *et al.*, 1998). The transfer of another class of MGEs, GTA, can also be induced by mitomycin C and oxidative stress (Stanton *et al.*, 2008).

Although it has been initially assumed that the SOS response is triggered exclusively by direct DNA damage, there is evidence that certain stimuli can indirectly generate SOS-inducing signals (Aertsen and Michiels, 2006). Some antibiotics, which do not directly interfere with DNA metabolism, may, nevertheless, induce a genuine SOS response. This was demonstrated, for example, for the β-lactam antibiotics targeting bacterial cell wall (Miller *et al.*, 2004). The resulting effect is the elevated rate of horizontal transfer of virulence genes in staphylococci (Maiques *et al.*, 2006; Ubeda *et al.*, 2006). Furthermore, the presence of a functional SOS response system seems not an absolute prerequisite for the stress-induced increase in HGT frequencies. In naturally competent bacterial species such as *Streptococcus pneumoniae* the antibiotic-imposed stress induces genetic transformation in the absence of a SOS-like system (Prudhomme *et al.*, 2006). Another naturally competent bacterium, *Legionella pneumophila,* also lacks a prototypic SOS response system, but UV light, which represents a major source of genotoxic stress in the environment, effectively induces competence development in this bacterium (Charpentier *et al.*, 2011). The authors have hypothesized that competence for natural transformation and, therefore, the ability to acquire and propagate foreign genes may have evolved as a DNA damage response in SOS-deficient bacteria. Thus, Gram-positive bacteria respond to stress conditions by induction of competence for genetic transformation to generate genetic diversity (Claverys *et al.*, 2006). The strategy is combined with SOS induction in bacteria such as *B. subtilis,* while others, such as *S. pneumoniae,* rely solely on competence.

In naturally competent Gram-negative bacteria, stress conditions may also enhance HGT. *V. cholerae,* for example, can acquire new genetic material by natural transformation during growth on chitin, which activates regulatory cascades leading to increased cell density, nutrient limitation, and decline in growth rate and stress (Meibom *et al.*, 2005). Stress in the form of DNA damage induces transcription and translation of competence genes in *Helicobacter pylori,* thus increasing transformation frequency and genetic exchange rates (Dorer *et al.*, 2010).

To what extent is the SOS response induced in natural ecosystems? The impact of solar UV radiation on microbiota may be profound, especially in aquatic ecosystems (Hader, 2000; Sinha and Hader, 2002). One of the major targets is DNA where UV-B irradiation results in the formation of cyclobutane-pyrimidine dimers (CPDs), 6-4 photoproducts (6-4PPs), and DNA strand breaks, thus leading to the induction of a number of repair mechanisms, including the SOS response (Rastogi *et al.*, 2010). In photosynthetic cyanobacteria, UV-B stress is accompanied by additional stresses such as oxidative stress and oxidative damage (He *et al.*, 2002). Despite the fact that future trends for solar UV irradiation of the Earth's surface remain unclear, the increases in UV-B irradiance over the latter part of the 20th century have been larger than the natural variability (McKenzie *et al.*, 2007). Thus, this factor may play an increasingly important role in accelerating HGT in microbial ecosystems.

Another important factor that may induce the SOS response in environmental bacteria is the release of DNA-damaging compounds as a result of human activities. Hospitals, for example, represent an indisputable release source of many chemical compounds in their wastewaters, with a significant genotoxic impact (Jolibois *et al.*, 2003; Jolibois and Guerbet, 2006, Gupta *et al.*, 2009). Genotoxic effects are also associated with wastewater of industrial and domestic origin (Jolibois and Guerbet, 2005; Isidori *et al.*, 2007). Then the release of the genotoxic wastewater into the environment, for example in the form of land application, may lead to the elevated genotoxicity in the corresponding ecosystems (Aleem and Malik, 2003; Alam *et al.*, 2009). In general, soil in highly industrialized areas displays elevated genotoxic effects (Xiao *et al.*, 2006). Genotoxicity or mutagenicity of environmental compounds is usually studied in regards to human and animal health but the techniques used to measure these effects such as SOS chromotest, umu test, Ames test and others are mostly based on bacterial systems. Thus, these measurements indirectly benefit our understanding of how the environmental compounds may affect HGT in environmental microbiota.

Antimicrobials

In recent years the concept of antibiotics as solely killing agents has been substantially reassessed in the light of new findings suggesting that low concentrations of antibiotics may play a regulatory function in natural ecosystems (Davies *et al.*, 2006; Fajardo and Martinez, 2008; Aminov, 2009). Moreover, the killing/therapeutic and regulatory/subinhibitory concentrations of an antibiotic are indeed directed towards different targets in the cell (Hoffman *et al.*, 2005). There is a substantial body of evidence suggesting that the subinhibitory concentrations of antibiotics may significantly increase the frequency of horizontal transfer of many types of mobile genetic elements. It was noticed a quarter of a century ago that the subinhibitory concentrations of β-lactams enhanced the transfer of tetracycline resistance plasmids in *S. aureus* by up to 1000-fold (Barr *et al.*, 1986). Pre-incubation of donor cells of *Bacteroides* sp. in the presence of subinhibitory tetracycline accelerates the mobilization of a resident non-conjugative plasmid by chromosomally encoded tetracycline conjugal elements (Valentine *et al.*, 1988). A similar type of exposure of donor *Bacteroides* cells appeared to be a pre-requisite for the excision and conjugal transfer of the ICE CTnDOT (Stevens *et al.*, 1993; Whittle *et al.*, 2002). In the absence of tetracycline induction of donor cells, practically no such transfers were detected. Addition of subinhibitory tetracycline into the mating medium substantially enhanced the conjugal transfer of another ICE, Tn*916* (Showsh and Andrews, 1992). A similar stimulatory effect of tetracycline on conjugation transfer was demonstrated for Tn*925* as well (Torres *et al.*, 1991). A recent study has revealed the induction effect of carbadox and metronidazole on the GTA-mediated HGT in *Brachyspira hyodysenteriae* (Stanton *et al.*, 2008). The induced VSH-1 particles transmitted tylosin and chloramphenicol resistances between *B. hyodysenteriae* strains.

These experiments described above have been performed under laboratory conditions using standard mating techniques, and the question is whether these observations reflect the real *in vivo* situation or these are the effects of specific *in vitro* conditions. The former point of view is supported by a number of experiments using animal models. For instance, the inclusion of subinhibitory tetracycline in drinking water resulted in a 10-fold increase of transfer of an ICE, Tn*1545*, from *Enterococcus faecalis* to *Listeria monocytogenes* in the intestine of gnotobiotic mice (Doucet-Populaire *et al.*, 1991). In gnotobiotic rats, the presence of tetracycline resulted in a higher number of Tn*916* transconjugants compared to control (Bahl *et al.*, 2004). An astonishing 100% transfer rate of a small plasmid pLFE1 from *Lactobacillus plantarum* to *E. faecalis* was observed in the gut of gnotobiotic mice receiving erythromycin (Feld *et al.*, 2008). These experiments suggest that the stimulatory effect of subinhibitory antibiotics on transfer of MGEs is not an artefact of *in vitro* conditions but happens in the real gut of real animals. It needs to be noted, however, that the transfer frequencies were estimated in gnotobiotic animals, which are lacking the highly diverse and dense microbiota in the gut.

The mechanisms contributing to the enhanced movement of MGEs in the presence of antibiotics have been established on several occasions, especially for ICEs. In Tn*916*, transcriptional regulation of the *tra* genes required for the conjugal transfer of this ICE is under the control of *orf7* and *orf8* products (Celli and Trieu-Cuot, 1998; Roberts and Mullany, 2009). Transcripts for these two ORFs are produced from the distant promoter of the *tet*(M) gene as well as from the promoter directly upstream of *orf7*. In the absence of tetracycline, transcription from P_{tetM} is attenuated (Su *et al.*, 1992), and the transcripts are very short, not covering the *orf7* and *orf8* genes. In addition, the product of *orf9* negatively regulates the promoter upstream of *orf7*. In the presence of tetracycline, however, transcription from P_{tetM} extends through *orf7* and *orf8* thus allowing the synthesis of these two proteins that promote transcription from P_{orf7}. The long transcripts from P_{tetM} are also complementary to the *orf9* transcripts thus efficiently reducing the concentration of P_{orf7} repressor. As a result, transcription from P_{orf7} extends downstream of *orf8* through *int* and *xis*, which encode the integrase and excisionase, respectively. Once Tn*916* is circularized, transcription from P_{orf7} also leads to the expression of the *tra* genes thus initiating the conjugation machinery.

Another mechanism, also involving tetracycline as a positive regulator of horizontal gene transfer, is implemented in the *Bacteroides* ICE, CTnDOT (Whittle *et al.*, 2002; Moon *et al.*, 2005). In this regulatory cascade, tetracycline activates the transcription of the *tet*(Q) gene and the downstream genes of the two-component regulatory system, *rteA* and *rteB*. The product of *rteB* activates the transcription of *rteC* system, which, in turn, leads to the elevated transcription of the gene cluster involved in CTnDOT excision. Subsequently, the proteins encoded by the excision genes upregulate the production of *tra* gene mRNA thus activating the conjugation machinery (Jeters *et al.*, 2009). In this complex regulatory system, the stimulatory effect of tetracycline on transcription of its own resistance gene (and concomitantly on the excision and conjugal transfer genes of CTnDOT) is implemented through a translation attenuation mechanism involving the leader region of *tet*(Q) (Wang *et al.*, 2005).

Both regulatory mechanisms of the MGE movement discussed above display a common theme in that the primary switch for this genetic process is based on an antibiotic-antibiotic resistance gene pair. That is, the presence of a subinhibitory antibiotic in the environment activates the transcription of a corresponding resistance gene and, concurrently, the genes involved in the mobility of MGEs. In this regard, the interaction of an antibiotic and antibiotic resistance gene resembles a positively regulated switch, with an antibiotic possessing a signalling function, ultimately leading to the activation of horizontal gene exchange in the microbiota. The recent works describing concentration-dependent bacterial responses to antibiotics have led to the development of the hormesis concept (Davies *et al.*, 2006; Fajardo and Martinez, 2008). According to this concept, low concentrations of antibiotics may regulate a specific set of genes in target bacteria, while increasingly higher concentrations elicit a stress response, and even higher concentrations are lethal. It has been suggested that antibiotics play a regulatory role in nature at low concentrations unlike the lethal concentrations used in clinical therapy (Aminov, 2009). Given the profound stimulatory effect of low-dose antibiotics on the movement of MGEs, one of the functions of antibiotics in natural ecosystems may be the regulation of HGT between the representatives of environmental microbiota.

Despite the fact that some countries have enacted legislations limiting the non-therapeutic use of antibiotics, in particularly in food animals (http://ec.europa.eu/food/food/animalnutrition/feedadditives/index_en.htm), the use of subtherapeutic antimicrobials in agriculture and aquaculture of other countries is still widespread. Moreover, the US Food and Drug Administration has recently approved subinhibitory concentrations of doxycycline and minocycline for the systemic treatment of skin infections in humans (Del Rosso, 2007). Thus, there is a high probability that the gut ecosystems may continue to be hotspots of horizontal gene exchange involving the resident and transient gut microbiota. The land application of manure with residual antibiotics and antibiotic resistance genes of the gut content may further contribute to the enhanced HGT in the environment (Chee-Sanford *et al.*, 2009).

Quorum sensing

The quorum sensing (QS) systems are widespread among a variety of microbiota and initially they were recognized as population-density-sensing mechanisms based on the best-studied prototype, the QS network, which regulates the *lux* operon in *Vibrio fischeri* (Eberhard *et al.*, 1981; Fuqua *et al.*, 1996). Since then it has become clear that the QS is involved in regulation of a much broader range of functions and activities such as pathogenicity, extracellular enzyme production, antibiotic biosynthesis, and others (Bainton *et al.*, 1992; Jones *et al.*, 1993; Passador *et al.*, 1993; Pirhonen *et al.*, 1993; Pierson *et al.*, 1994). The QS is indeed a universal language of communication not only among the bacteria but also in inter-kingdom interaction (Shiner *et al.*, 2005).

One of the earliest indications of the QS involvement in HGT came from the studies of conjugal transfer of the *Agrobacterium tumefaciens* Ti plasmids (Zhang and Kerr, 1991; Piper *et al.*, 1993; Fuqua and Winans, 1994). The Ti plasmid encodes a regulatory system, consisting of the acylhomoserine lactone (AHL) synthase TraI and the transcription factor TraR. TraI synthesizes AHLs, mainly N-(3-oxo-octanoyl)-L-homoserine lactone (OOHL), while TraR is an OOHL-dependent transcription activator of conjugative transfer genes. Molecular mechanisms of this activation that leads to the enhanced conjugal transfer of the Ti plasmids are well understood (Costa *et al.*, 2009; Quin *et al.*, 2009).

Another well-explored area is the QS regulation of transfer of large symbiotic plasmids in rhizobia (Danino *et al.*, 2003; He *et al.*, 2003; Tun-Garrido *et al.*, 2003). Analysis of plasmid transfer in several rhizobia species has revealed a regulatory relay that is specifically poised to detect AHLs made by different cells and to respond to these signals by up-regulation of conjugal transfer genes. In turn, the production of AHLs, N-(3-oxo-octanoyl)-L-homoserine lactone and N-(octanoyl)-L-homoserine lactone, is regulated by a complex interaction of plasmid- and chromosome-encoded genes in donors and recipients in response to environmental cues. Recently it has been shown that the transfer of an ICE of *Mesorhizobium loti*, which carries genes for a nitrogen–fixing symbiosis with *Lotus* species, is also regulated by a QS mechanism (Ramsay *et al.*, 2009). The two conserved hypothetical genes, which are essential for the QS-mediated excision and transfer, can also be found on putative ICEs in several alpha-proteobacteria, indicating a broader presence of this HGT regulatory mechanism.

Recent investigations have identified an important role played by QS in regulation of phage-mediated HGT. For example, acyl-homoserine lactones (AHLs), the essential signalling molecules of QS in many Gram-negative bacteria, can trigger phage production in soil and groundwater bacteria (Ghosh *et al.*, 2009). Interestingly, in the *recA* mutant of *E. coli* the induction responses of lambda to AHL remained unaffected, suggesting that this mechanism does not involve a SOS response.

Genetic mechanisms involved in maintenance and dissemination of MGEs

The 'cooperation' between different mechanisms of HGT can be seen in many examples of mobility among the genetic elements that are normally not mobile on their own. For example, ICEs of *Bacteroides* spp. and large broad-host-range conjugative IncP plasmids R751 and RP4 are able to act *in trans* to excise, circularize, and transfer unlinked integrated elements called NBUs (non-replicating *Bacteroides* units) (Li *et al.*, 1993; Shoemaker *et al.*, 1993). Bioinformatics of plasmid mobility suggests that globally about a quarter of all plasmids are potentially mobilizable and, therefore, can be transferred if a compatible conjugation machinery is present (Smillie *et al.*, 2010). In *Sinorhizobium meliloti*, conjugation functions for a 1683 kb symbiotic megaplasmid pSymB are supplied *in trans* by another megaplasmid, pSymB (Blanca-Ordóñez *et al.*, 2010). The recently described IS*CR* elements are thought to be the key players in IncA/C plasmid evolution serving as antibiotic resistance gene capture and movement systems that are also capable of constructing extended clusters of antibiotic resistance genes (Toleman and Walsh, 2010).

Insertion sequences (ISs) constitute an important component of most bacterial genomes and are simple transposable elements consisting of inverted repeat (IR) sequences, a transposase gene, and frequently a second recombination regulation enzyme gene (Mahillon and Chandler, 1998). It has been thought that the transposition of ISs is a rare event within a bacterial genome, but the discovery of a protein called IS-excision enhancer (IEE), which promotes excision events, suggested that the rates could be high (Kusumoto *et al.*, 2011). The IEE activity, therefore, may play an important role in bacterial genome evolution by inducing IS removal and genomic deletion. Another aspect is the transposition of an IS into a conjugative plasmid, an ICE, or a genomic island. This way, the IS elements can be disseminated to many other, not necessarily closely related, bacterial taxa. Together with prophages (Asadulghani *et al.*, 2009), IS elements are the major contributors to the genomic diversification in pathogenic *E. coli* (Ooka *et al.*, 2009). Since the IS elements are ubiquitous in many bacteria, including environmental and commensal species, these mechanisms of diversification are probably not limited exclusively to pathogens.

Since the carriage of MGEs may be associated with a considerable fitness cost, certain mechanisms of stable inheritance have been selected within the mobile genetic elements. One of these mechanisms is a toxin–antitoxin (TA) system consisting of two components, a stable toxin and its labile antitoxin. Type I and II TA systems were found on plasmids in the 1980s (Ogura and Higara, 1983; Jaffe *et al.*, 1985), and type III is a more recent discovery (Fineran *et al.*, 2009). The general mechanism of action is that if an MGE with TA system is lost during cell division, the concentration of a labile antitoxin rapidly decreases thus releasing a stable toxin, which kills an MGE-free cell (Van Melderen, 2010). Thus, TA systems contribute to the stable maintenance and dissemination of plasmids and genomic islands in bacterial populations despite the associated fitness cost.

The size of MGEs varies from small IS elements (typically 700 to 2500 bp) (Mahillon and Chandler, 1998) to large symbiotic megaplasmids of *Sinorhizobium meliloti* (1.35 Mb and 1.68 Mb) (Barnett *et al.*, 2001; Capela *et al.*, 2001). The majority of large self-transmissible genetic elements have a sufficient capacity to carry multiple genes, including those encoding antibiotic, heavy metal and biocide resistances, metabolism of various substrates and xenobiotics, symbiosis with the host, and other auxiliary functions. The fact that R plasmids mediate resistance to mercury, nickel and cobalt was first described more than 40 years ago (Smith, 1967). Subsequent research by other groups demonstrated that the genetic linkage between antibiotic resistance and mercury resistance in enterobacteria had occurred prior to the late 1950s in Japan (reviewed in Liebert *et al.*, 1999). At the same time, Enterobacteriaceae strains collected by E.D.G. Murray from 1917 to 1954 contained very few antibiotic and mercury resistant bacteria, despite the finding that 25% of the strains carried conjugative functions (Hughes and Datta, 1983). Thus, the frequency of MGEs even in the 'pre-antibiotic era' was sufficiently high but not associated with the adaptation to anthropogenic factors such as antibiotics or heavy metals.

The exceptional natural genetic engineering capabilities of bacteria have been profoundly demonstrated during the 'antibiotic era' (Aminov, 2010). To withstand the massive pressure of antimicrobials used by humans, commensal and pathogenic bacteria were able, within a relatively short period of time on the evolutionary scale, to mobilize a huge reservoir of antibiotic resistance genes, often from the environmental bacteria (Aminov and Mackie, 2007; Cantón, 2009; Wright, 2010). The main genetic engineering tool used by bacteria to collect antibiotic resistance genes is integrons, the genetic platform that is involved in the acquisition and functional expression of exogenous gene cassettes (Mazel, 2006). The discovery of superintegrons that contain hundreds of auxiliary genes and may occupy a significant part of many bacterial genomes has changed our initial interpretation of integrons as merely a mechanism of collecting antibiotic resistance genes. The presence of toxin/antitoxin cassettes in superintegrons further facilitates the stabilization of large cassette arrays consisting of many ancillary genes (Cambray *et al.*, 2010).

The physical linkage of numerous and functionally diverse groups of genes within the high-capacity MGEs has implications for their persistence in the environment. Selection that is imposed towards even a single component/gene of an MGE will automatically select for the whole MGE. Thus, we should see the co-selection effect in phenotypes of bacteria that carry MGEs. Indeed, bacteria in metal-contaminated areas appeared to be more tolerant to metals and antibiotics than in control sites (Stepanauskas *et al.*, 2005; Baker-Austin *et al.*, 2006; Wright *et al.*, 2006). Treatment of agricultural soils with copper may lead to a significantly higher incidence of antibiotic resistance phenotypes in indigenous soil microbiota (Berg *et al.*, 2005). In freshwater microcosms, amendment with metal concentrations representative of industry and mining-impacted environments increased the frequency of antibiotic resistance in the microbial communities (Stepanauskas *et al.*, 2006). Genetic mechanisms responsible for the co-selection phenomena in the contaminated environments are presently poorly understood, but it is clear that MGE-driven HGT is the main adaptive trait in bacteria inhabiting industrially contaminated aquatic (Wright *et al.*, 2008) and soil (Top *et al.*, 1995; Sobecky and Coombs, 2009) ecosystems.

On the other side of the spectrum are the negative regulators of HGT, genetic mechanisms that constrain gene transfer events. Some of the best-studied systems are receptor masking, restriction-modification and abortive infection that protect the cell from foreign DNA (Hyman and Abedon, 2010; Labrie *et al.*, 2010). This is especially important if it is a bacteriophage DNA that may enter into the lytic cycle and thus destroy the bacterial host. The previously discussed stress and SOS responses are involved into the regulation of restriction-modification activities. Alleviation of EcoK DNA restriction in *E. coli* by UV irradiation and expression of the SOS response, for example, results in a 10^3- to 10^4-fold increase in phage infection and in a 4-fold increase in plasmid transformation rates (Hiom and Sedgwick, 1992). All four resident restriction systems of *E. coli* K-12 show reduced activity following UV treatment (Kelleher and Raleigh, 1994). Thus the previously discussed effects of the SOS response induction also involve the suppression of restriction activity leading to the increased rate of phage infection and natural transformation.

Clustered regularly interspaced short palindromic repeats (CRISPRs) in the genomic sequences of Bacteria and Archaea were discovered during *in silico* analyses almost a decade ago (Jansen *et al.*, 2002). These are direct repeats, varying in size from 21 to 50 bp, interspaced by similarly sized non-repetitive sequences. Initially, four CRISPR-associated (cas) genes were identified in CRISPR-containing prokaryotes that were absent from

CRISPR-negative prokaryotes. Further analyses showed that up to 45 CRISPR-associated (Cas) protein families are uniformly associated with CRISPRs and located near a repeat cluster (Haft *et al.*, 2005). The biological role of CRISPRs remained elusive until 2007, when the experimental evidence suggested that these elements provide acquired resistance against viruses (Barrangou *et al.*, 2007). Following viral challenge, bacteria integrate new spacers derived from the phage genome and resistance specificity is determined by spacer-phage sequence similarity. Further investigations revealed that this CRISPR/Cas bacterial immune system cleaves invading bacteriophage and plasmid DNA in a manner similar to the RNA interference mechanism in eukaryotes (Garneau *et al.*, 2010).

Very little is known about the regulation of CRISPR/Cas systems. In *E. coli* K12 transcription from the *cas*A and CRISPR I promoters is repressed by a heat-stable nucleoid-structuring protein (H-NS) (Pul *et al.*, 2010), which is a global transcription repressor in Gram-negative bacteria. This repression can be relieved by a number of regulators, in particular by LeuO (Cen and Wu, 2005, Westra *et al.*, 2010), a member of the LysR family of transcription factors in Gram-negative enteric bacteria (Stoebel *et al.*, 2008). In *Salmonella enterica* serovar Typhi the CRISPR/Cas system appeared to be regulated similarly, by the antagonistic activities of LeuO and H-NS (Medina-Aparicio *et al.*, 2011). In general, however, the CRISPR/Cas system in wild-type *E. coli* is repressed under laboratory conditions and it does not provide efficient protection against phages (Westra *et al.*, 2010). The involvement of a broader range of model organisms as well as exploration of other cellular regulatory mechanisms may provide a clue on how the CRISPR-based immunity is regulated by environmental factors.

Concluding remarks

It is hypothesized that the extent of horizontal gene exchange in the early stages of life on the Earth was much higher, dominating the evolutionary landscape (Woese, 2000). Only after evolving into more integrated and stable cellular entities the true organismal lineages in the Darwinian sense (e.g. evolving by vertical evolution that is enacted through the natural selection of the fittest among genetically variable individuals) could exist. But, even after the establishment of the three domains of life, HGT continued to play an important role in evolution. The possibility of lateral acquisition of a gene or a group of genes that may confer immediate selective advantage substantially accelerates the evolutionary process; the situation with rapid dissemination of antibiotic resistance serves as a perfect example of this scenario.

It is evident from retrospective and prospective studies that the rate of horizontal gene exchange in the microbial world is high indeed. The remnants of the past horizontal gene acquisition can be clearly seen in all microbial genomes sequenced to date. It is a sort of ironic situation to realize that many phenotypic characteristics used in microbial taxonomy for years in defining the identity of species, such as the range of substrates utilized, were actually acquired from other organisms. The microbial world around us can be seen as a gigantic microbiome, with the continuous flow of genes between its different compartments. This flow is sustained by a variety of sophisticated natural genetic engineering tools, MGEs, which have been selected during evolution as providing the means for re-shuffling the available genetic material and picking up the best responses possible to cope with the continuously changing environmental challenges. The recent relatively short history of the 'antibiotic era' (Aminov, 2010) demonstrates the ultimate success of this strategy and urges us to rethink our own when interacting with the microbial world.

Continuous discoveries of novel MGEs and mechanisms of HGT, together with the findings of unexpectedly high HGT rates in natural ecosystems, indicate that we are still far from understanding the true extent of HGT in nature. A contribution to better understanding may be envisaged as the combination of retrospective and prospective approaches. On the dry lab side, the history of past HGT events, which is recorded in the wealth of genomic information, can be more vigorously interrogated on the basis of our knowledge about MGEs and with the help of bioinformatics tools that are able to detect the events consistent with HGT. On the wet lab side, the need is the development of *in situ* technologies that are more sensitive, less intrusive and applicable to field studies. Microcosm experiments should model real environmental situations as close as possible, working with native microbiota, with a lesser dependence on model organisms. These developments may help to elaborate better strategies to deal with the pressing needs such as the emergence of novel infections and opportunistic pathogens as well as antibiotic resistance genes.

Acknowledgements

The author acknowledges the support of the Rural and Environmental Science and Analytical Services (RESAS) Programme of the Scottish Government and the Department for Environment, Food and Rural Affairs (DEFRA, project OD2014).

References

Aardema, B.W., Lorenz, M.G., and Krumbein, W.E. (1983). Protection of sediment-adsorbed transforming DNA against enzymatic inactivation. Appl. Environ. Microbiol. *46*, 417–420.

Aertsen, A., and Michiels, C.W. (2006). Upstream of the SOS response: figure out the trigger. Trends Microbiol. *14*, 421–423.

Akhtar, M., Hirt, H., and Zurek, L. (2009). Horizontal transfer of the tetracycline resistance gene *tet*(M) mediated by pCF10 among *Enterococcus faecalis* in the house fly (*Musca domestica* L.) alimentary canal. Microb. Ecol. *58*, 509–518.

Akkermans, A.D.L., Van Elsas, J.D., and De Bruijn, F.J. (1995). Molecular Microbial Ecology Manual (Kluwer Academic Publisher, Dordrecht, the Netherlands).

Alam, M.Z., Ahmad, S., and Malik, A. (2009). Genotoxic and mutagenic potential of agricultural soil irrigated with tannery effluents at Jajmau (Kanpur), India. Arch. Environ. Contam. Toxicol. *57*, 463–476.

Aleem, A., and Malik, A. (2003). Genotoxic hazards of long-term application of wastewater on agricultural soil. Mutat. Res. *538*, 145–154.

Allen, M.A., Lauro, F.M., Williams, T.J., Burg, D., Siddiqui, K.S., De Francisci, D., Chong, K.W., Pilak, O., Chew, H.H., De Maere, M.Z., *et al.* (2009). The genome sequence of the psychrophilic archaeon, *Methanococcoides burtonii*: the role of genome evolution in cold adaptation. ISME J. *3*, 1012–1035.

Aminov, R.I. (2009). The role of antibiotics and antibiotic resistance in nature. Environ. Microbiol. *11*, 2970–2988.

Aminov, R.I. (2010). A brief history of the antibiotic era: lessons learned and challenges for the future. Front. Microbiol. *1*, 134. doi:10.3389/fmicb.2010.00134.

Aminov, R.I., and Mackie, R.I. (2007). Evolution and ecology of antibiotic resistance genes. FEMS Microbiol. Lett. *271*, 147–161.

Anagnostopoulos, C., and Spizizen, J. (1961). Requirements for transformation in *Bacillus subtilis.* J. Bacteriol. *81*, 741–746.

Andersen, J.B., Sternberg, C., Poulsen, L.K., Bjorn, S.P., Givskov, M., and Molin, S. (1998). New unstable variants of green fluorescent protein for studies of transient gene expression in bacteria. Appl. Environ. Microbiol. *64*, 2240–2246.

Anderson, G.G., and O'Toole, G.A. (2008). Innate and induced resistance mechanisms of bacterial biofilms. Curr. Top Microbiol. Immunol. *322,* 85–105.

Angles, M.L., Marshall, K.C., and Goodman, A.E. (1993). Plasmid transfer between marine bacteria in the aqueous phase and biofilms in reactor microcosms. Appl. Environ. Microbiol. *59,* 843–850.

Angly, F.E., Felts, B., Breitbart, M., Salamon, P., Edwards, R.A., Carlson, C., Chan, A.M., Haynes, M., Kelley, S., Liu, H., *et al.* (2006). The marine viromes of four oceanic regions. PLoS Biol. *4,* e368.

Asadulghani, M., Ogura, Y., Ooka, T., Itoh, T., Sawaguchi, A., Iguchi, A., Nakayama, K., and Hayashi, T. (2009). The defective prophage pool of *Escherichia coli* O157: prophage–prophage interactions potentiate horizontal transfer of virulence determinants. PLoS Pathog. *5,* e1000408.

Auchtung, J.M., Lee, C.A., Monson, R.E., Lehman, A.P., and Grossman, A.D. (2005). Regulation of a *Bacillus subtilis* mobile genetic element by intercellular signaling and the global DNA damage response. Proc. Natl. Acad. Sci. U.S.A. *102,* 12554–12559.

Averhoff, B., and Friedrich, A. (2003). Type IV pili-related natural transformation systems: DNA transport in mesophilic and thermophilic bacteria. Arch. Microbiol. *180,* 385–393.

Avery, O.T., MacLeod C.M., and McCarty, M. (1944). Studies on the chemical nature of the substance inducing transformation of pneumococcal types. Induction of transformation by a desoxyribonucleic acid fraction isolated from pneumococcus type III. J. Exp. Med. *89,* 137–158.

Bahar, O., Goffer, T., and Burdman, S. (2009). Type IV pili are required for virulence, twitching motility, and biofilm formation of *Acidovorax avenae* subsp. *citrulli*. Mol. Plant Microbe Interact. *22,* 909–920.

Bahl, M.I., Sorensen, S.J., Hansen, L.H., and Licht, T.R. (2004). Effect of tetracycline on transfer and establishment of the tetracycline-inducible conjugative transposon Tn*916* in the guts of gnotobiotic rats. Appl. Environ. Microbiol. *70,* 758–764.

Bainton, N.J., Bycroft, B.W., Chhabra, S.R., Stead, P., Gledhill, L., Hill, P.J., Rees, C.E., Winson, M.K., Salmond, G.P., Stewart, G.S., *et al.* (1992). A general role for the lux autoinducer in bacterial cell signaling: control of antibiotic biosynthesis in *Erwinia*. Gene *116,* 87–91.

Baker-Austin, C., Wright, M.S., Stepanauskas, R., and McArthur, J.V. (2006). Co-selection of antibiotic and metal resistance. Trends Microbiol. *14,* 176–182.

Bamford C.V., d'Mello, A., Nobbs, A.H., Dutton, L.C., Vickerman, M.M., and Jenkinson, H.F. (2009). *Streptococcus gordonii* modulates *Candida albicans* biofilm formation through intergeneric communication. Infect. Immun. *77,* 3696–3704.

Barken, K.B., Pamp, S.J., Yang, L., Gjermansen, M., Bertrand, J.J., Klausen, M., Givskov, M., Whitchurch, C.B., Engel, J.N., and Tolker-Nielsen, T. (2008). Roles of type IV pili, flagellum-mediated motility and extracellular DNA in the formation of mature multicellular structures in *Pseudomonas aeruginosa* biofilms. Environ. Microbiol. *10,* 2331–2343.

Barnett, M.J., Fisher, R.F., Jones, T., Komp, C., Abola, A.P., Barloy-Hubler, F., Bowser, L., Capela, D., Galibert, F., Gouzy, J., *et al.* (2001). Nucleotide sequence and predicted functions of the entire *Sinorhizobium meliloti* pSymA megaplasmid. Proc. Natl. Acad. Sci. U.S.A. *98,* 9883–9888.

Barondess, J.J., and Beckwith, J. (1990). A bacterial virulence determinant encoded by lysogenic coliphage lambda. Nature *346,* 871–874.

Barr, V., Barr, K., Millar, M.R., and Lacey, R.W. (1986). Beta-lactam antibiotics increase the frequency of plasmid transfer in *Staphylococcus aureus*. J. Antimicrob. Chemother. 17: 409–413.

Barrangou, R., Fremaux, C., Deveau, H., Richards, M., Boyaval, P., Moineau, S., Romero, D.A., and Horvath, P. (2007). CRISPR provides acquired resistance against viruses in prokaryotes. Science *315,* 1709–1712.

Baya, A.M., Brayton, P.R., Brown, V.L., Grimes, D.J., Russek-Cohen, E., and Colwell, R.R. (1986). Coincident plasmids and antimicrobial resistance in marine bacteria isolated from polluted and unpolluted Atlantic Ocean samples. Appl. Environ. Microbiol. *51,* 1285–1292.

Beaber, J.W., Hochhut, B., and Waldor, M.K. (2004). SOS response promotes horizontal dissemination of antibiotic resistance genes. Nature *427,* 72–74.

Berg, J., Tom-Petersen, A., and Nybroe, O. (2005). Copper amendment of agricultural soil selects for bacterial antibiotic resistance in the field. Lett. Appl. Mircobiol. *40,* 146–151.

Binh, C.T., Heuer, H., Kaupenjohann, M., and Smalla, K. (2008). Piggery manure used for soil fertilization is a reservoir for transferable antibiotic resistance plasmids. FEMS Mircobiol. Ecol. *66,* 25–37.

Binh, C.T., Heuer, H., Kaupenjohann, M., and Smalla, K. (2009). Diverse aadA gene cassettes on class 1 integrons introduced into soil via spread manure. Res. Microbiol. *160,* 427–433.

Björklöf, K., Suoniemi, A., Haahtela, K., and Romantschuk, M. (1995). High frequency of conjugation versus plasmid segregation of RP1 in epiphytic *Pseudomonas syringae* populations. Microbiology *141,* 2719–2727.

Blanca-Ordóñez, H., Oliva-García, J.J., Pérez-Mendoza, D., Soto, M.J., Olivares, J., Sanjuán, J., and Nogales, J. (2010). pSymA-dependent mobilization of the *Sinorhizobium meliloti* pSymB megaplasmid. J. Bacteriol. *192*, 6309–6312.

Bose, B., Auchtung, J.M., Lee, C.A., and Grossman, A.D. (2008). A conserved anti-repressor controls horizontal gene transfer by proteolysis. Mol. Microbiol. *70*, 570–582.

Bosó-Ribelles, V., Romá-Sánchez, E., Carmena, J., Cáceres, C., and Bautista, D. (2008). Tigecycline: a new treatment choice against *Acinetobacter baumannii*. Recent Pat. Antiinfect. Drug Discov. *3*, 117–122.

Boucher, Y., Cordero, O.X., Takemura, A., Hunt, D.E., Schliep, K., Bapteste, E., Lopez, P., Tarr, C.L., and Polz, M.F. (2011). Local mobile gene pools rapidly cross species boundaries to create endemicity within global *Vibrio cholerae* populations. mBio *2*, e00335–10.

Brabban, A.D., Hite, E., and Callaway, T.R. (2005). Evolution of foodborne pathogens via temperate bacteriophage-mediated gene transfer. Foodborne Pathog. Dis. *2*, 287–303.

Breitbart, M., Hewson, I., Felts, B., Mahaffy, J.M., Nulton, J., Salamon, P., and Rohwer, F. (2003). Metagenomic analyses of an uncultured viral community from human feces. J. Bacteriol. *185*, 6220–6223.

Brigulla, M., and Wackernagel, W. (2010). Molecular aspects of gene transfer and foreign DNA acquisition in prokaryotes with regard to safety issues. Appl. Mircobiol. Biotechnol. *86*, 1027–1041.

Brinkhoff, T., Giebel, H.A., and Simon, M. (2008). Diversity, ecology, and genomics of the Roseobacter clade: a short overview. Arch. Microbiol. *189*, 531–539.

Brochier-Armanet, C., Deschamps, P., López-García, P., Zivanovic, Y., Rodríguez-Valera, F., and Moreira, D. (2011). Complete-fosmid and fosmid-end sequences reveal frequent horizontal gene transfers in marine uncultured planktonic archaea. ISME J. *5*, 1291–1302.

Buckling, A., and Rainey, P.B. (2002). The role of parasites in sympatric and allopatric host diversification. Nature *420*, 496–499.

Burmølle, M., Bahl, M.I., Jensen, L.B., Sørensen, S.J., and Hansen, L.H. (2008). Type 3 fimbriae, encoded by the conjugative plasmid pOLA52, enhance biofilm formation and transfer frequencies in Enterobacteriaceae strains. Microbiology *154*, 187–195.

Burroughs, N.J., Marsh, P., and Wellington, E.M. (2000). Mathematical analysis of growth and interaction dynamics of streptomycetes and a bacteriophage in soil. Appl. Environ. Microbiol. *66*, 3868–3877.

Burrus, V., Pavlovic, G., Decaris, B., and Guédon, G. (2002). Conjugative transposons: the tip of the iceberg. Mol. Microbiol. *46*, 601–610.

Cai, P., Huang, Q.Y., and Zhang, X.W. (2006). Interactions of DNA with clay minerals and soil colloidal particles and protection against degradation by DNase. Environ. Sci. Technol. *40*, 2971–2976.

Calderwood S.B., Auclair, F., Donohue-Rolfe, A., Keusch, G.T., and Mekalanos, J.J. (1987). Nucleotide sequence of the Shiga-like toxin genes of *Escherichia coli*. Proc. Natl. Acad. Sci. U.S.A. *84*, 4364–4368.

Cambray, G., Guerout, A.M., and Mazel, D. (2010). Integrons. Annu. Rev. Genet. *44*, 141–166.

Cantón, R. (2009). Antibiotic resistance genes from the environment: a perspective through newly identified antibiotic resistance mechanisms in the clinical setting. Clin. Mircobiol. Infect. *15(Suppl. 1)*, 20–25.

Capela, D., Barloy-Hubler, F., Gouzy, J., Bothe, G., Ampe, F., Batut, J., Boistard, P., Becker, A., Boutry, M., Cadieu, E., *et al.* (2001). Analysis of the chromosome sequence of the legume symbiont *Sinorhizobium meliloti* strain (1021). Proc. Natl. Acad. Sci. U.S.A. *98*, 9877–9882.

Caro-Quintero, A., Deng, J., Auchtung, J., Brettar, I., Höfle, M.G., Klappenbach, J., and Konstantinidis, K.T. (2011). Unprecedented levels of horizontal gene transfer among spatially co-occurring *Shewanella* bacteria from the Baltic Sea. ISME J. *5*, 131–140.

Carrolo, M., Frias, M.J., Pinto, F.R., Melo-Cristino, J., and Ramirez, M. (2010). Prophage spontaneous activation promotes DNA release enhancing biofilm formation in *Streptococcus pneumoniae*. PLoS One *5*, e15678.

Celli, J., and Trieu-Cuot, P. (1998). Circularization of Tn*916* is required for expression of the transposon-encoded transfer functions: characterization of long tetracycline-inducible transcripts reading through the attachment site. Mol. Microbiol. *28*, 103–117.

Charpentier, X., Kay, E., Schneider, D., and Shuman, H.A. (2011). Antibiotics and UV radiation induce competence for natural transformation in *Legionella pneumophila*. J. Bacteriol. *193*, 1114–1121.

Chee-Sanford J.C., Mackie, R.I., Koike, S., Krapac, I.G., Lin, Y.F., Yannarell, A.C., Maxwell, S., and Aminov, R.I. (2009). Fate and transport of antibiotic residues and antibiotic resistance genes following land application of manure waste. J. Environ. Qual. *38*, 1086–1108.

Chen, C.C., and Wu, H.Y. (2005). LeuO protein delimits the transcriptionally active and repressive domains on the bacterial chromosome. J. Biol. Chem. *280,* 15111–15121.

Christensen, B.B., Sternberg, C., and Molin, S. (1996). Bacterial plasmid conjugation on semi-solid surfaces monitored with the green fluorescent protein from *Aequorea victoria* as a marker. Gene *173,* 59–65.

Clarke, M.B., Hughes, D.T., Zhu, C., Boedeker, E.C., and Sperandio, V. (2006). The QseC sensor kinase: a bacterial adrenergic receptor. Proc. Natl. Acad. Sci. U.S.A. *103,* 10420–10425.

Claverys, J.P., Prudhomme, M., and Martin, B. (2006). Induction of competence regulons as a general response to stress in Gram-positive bacteria. Annu. Rev. Microbiol. *60,* 451–475.

Coombs, J.M., and Barkay, T. (2004). Molecular evidence for the evolution of metal homeostasis genes by lateral gene transfer in bacteria from the deep terrestrial subsurface. Appl. Environ. Microbiol. *70,* 1698–1707.

Costa, E.D., Cho, H., and Winans, S.C. (2009). Identification of amino acid residues of the pheromone-binding domain of the transcription factor TraR that are required for positive control. Mol. Microbiol. *73,* 341–351.

Costerton, J.W. (1995). Overview of microbial biofilms. J. Ind. Microbiol. *15,* 137–140.

Coombs, J.M. (2009). Potential for horizontal gene transfer in microbial communities of the terrestrial subsurface. Methods Mol. Biol. *532,* 413–433.

Daane, L.L., Molina, J.A., Berry, E.C., and Sadowsky, M.J. (1996). Influence of earthworm activity on gene transfer from *Pseudomonas fluorescens* to indigenous soil bacteria. Appl. Environ. Microbiol. *62,* 515–521.

Daane, L.L., Molina, J.A.E., and Sadowsky, M.J. (1997). Plasmid transfer between spatially separated donor and recipient bacteria in earthworm-containing soil microcosms. Appl. Environ. Microbiol. *63,* 679–686.

Dahl, K.H., Mater, D.D., Flores, M.J., Johnsen, P.J., Midtvedt, T., Corthier, G., and Sundsfjord, A. (2007). Transfer of plasmid and chromosomal glycopeptide resistance determinants occurs more readily in the digestive tract of mice than *in vitro* and exconjugants can persist stably *in vivo* in the absence of glycopeptide selection. J. Antimicrob. Chemother. *59,* 478–486.

Dahlberg, C., Bergström, M., Andreasen, M., Christensen, B.B., Molin, S., and Hermansson, M. (1998a). Interspecies bacterial conjugation by plasmids from marine environments visualized by gfp expression. Mol. Biol. Evol. *15,* 385–390.

Dahlberg, C., Bergström, M., and Hermansson, M. (1998b). In situ detection of high levels of horizontal plasmid transfer in marine bacterial communities. Appl. Environ. Microbiol. *64,* 2670–2675.

Damier-Piolle, L., Magnet, S., Brémont, S., Lambert, T., and Courvalin, P. (2008). AdeIJK, a resistance-nodulation-cell division pump effluxing multiple antibiotics in *Acinetobacter baumannii.* Antimicrob. Agents Chemother. *52,* 557–562.

Danino, V.E., Wilkinson, A., Edwards, A., and Downie, J.A. (2003). Recipient-induced transfer of the symbiotic plasmid pRL1JI in Rhizobium *leguminosarum* bv. *viciae* is regulated by a quorum-sensing relay. Mol. Microbiol. *50,* 511–512.

Datta, N., and Hedges, R.W. (1971). Compatibility groups among fi^- R factors. Nature *234,* 222–223.

Davey, M.E., and O'Toole, G.A. (2000). Microbial biofilms: from ecology to molecular genetics. Mircobiol. Mol. Biol. Rev. *64,* 847–867.

Davies, J., Spiegelman, G.B., and Yim, G. (2006). The world of subinhibitory antibiotic concentrations. Curr. Opin. Microbiol. *9,* 445–453.

Del Rosso, J.Q. (2007). Recently approved systemic therapies for acne vulgaris and rosacea. Cutis *80,* 113–120.

Demanèche, S., Jocteur-Monrozier, L., Quiquampoix, H., and Simonet, P. (2001). Evaluation of biological and physical protection against nuclease degradation of clay-bound plasmid DNA. Appl. Environ. Microbiol. *67,* 293–299.

Deni, J., Message, B., Chioccioli, M., and Tepfer, D. (2005). Unsuccessful search for DNA transfer from transgenic plants to bacteria in the intestine of the tobacco hornworm, *Manduca sexta.* Transgenic Res. *14,* 207–215.

Dickschat, J.S. (2010). Quorum sensing and bacterial biofilms. Nat. Prod. Rep. *27,* 343–369.

Dijkshoorn, L., Nemec, A., and Seifert, H. (2007). An increasing threat in hospitals: multidrug-resistant *Acinetobacter baumannii.* Nat. Rev. Microbiol. *5,* 939–951.

Dionisio, F., Matic, I., Radman, M., Rodrigues, O.R., and Taddei, F. (2002). Plasmids spread very fast in heterogeneous bacterial communities. Genetics *162,* 1525–1532.

Dorer, M.S., Fero, J., and Salama, N.R. (2010). DNA damage triggers genetic exchange in *Helicobacter pylori*. PLoS Pathog. *6*, e1001026.

Doucet-Populaire, F., Trieu-Cuot, P., Dosbaa, I., Andremont, A., and Courvalin, P. (1991). Inducible transfer of conjugative transposon Tn*1545* from *Enterococcus faecalis* to *Listeria monocytogenes* in the digestive tracts of gnotobiotic mice. Antimicrob. Agents Chemother. *35*, 185–187.

Duckworth, D.H. (1976). Who discovered bacteriophage? Bacteriol. Rev. *40*, 793–802.

Eberhard, A., Burlingame, A.L., Eberhard, C., Kenyon, G.L., Nealson, K.H., and Oppenheimer, N.J. (1981). Structural identification of autoinducer of *Photobacterium fischeri* luciferase. Biochemistry *20*, 2444–2449.

Edwards, R.A., and Rohwer, F. (2005). Viral metagenomics. Nat. Rev. Microbiol. *3*, 504–510.

Elkins, C.A., Moser, S.A., and Savage, D.C. (2001). Genes encoding bile salt hydrolases and conjugated bile salt transporters in *Lactobacillus johnsonii* 100–100 and other *Lactobacillus* species. Microbiology *147*, 3403–3412.

van Elsas, J.D., and Bailey, M.J. (2002). The ecology of transfer of mobile genetic elements. FEMS Mircobiol. Ecol. *42*, 187–197.

van Elsas, J.D., Trevors, J.T., and Starodub, M.E. (1988). Bacterial conjugation between pseudomonads in the rhizosphere of wheat. FEMS Microbiol. Ecol. *54*, 299–306.

van Elsas, J.D., Fry, J.C., Hirsch, P., and Molin, S. (2000). Ecology of plasmid transfer and spread. In The Horizontal Gene Pool; Bacterial Plasmids and Gene Spread, Thomas, C.M., ed. (Harwood Scientific Publisher, Amsterdam, the Netherlands), pp. 175–206.

Estrela, A.B., Heck, M.G., and Abraham, W.R. (2009). Novel approaches to control biofilm infections. Curr. Med. Chem. *16*, 1512–1530.

Etchuuya, R., Ito, M., Kitano, S., Shigi, F., Sobue, R., and Maeda, S. (2011). Cell-to-cell transformation in *Escherichia coli*: a novel type of natural transformation involving cell-derived DNA and a putative promoting pheromone. PLoS One *6*, e16355.

Fajardo, A., and Martínez, J.L. (2008). Antibiotics as signals that trigger specific bacterial responses. Curr. Opin. Microbiol. *11*, 161–167.

Falsetta, M.L., Bair, T.B., Ku, S.C., Vanden Hoven, R.N., Steichen, C.T., McEwan, A.G., Jennings, M.P., and Apicella, M.A. (2009). Transcriptional profiling identifies the metabolic phenotype of gonococcal biofilms. Infect. Immun. *77*, 3522–3532.

Feld L., Schjørring, S., Hammer, K., Licht, T.R., Danielsen, M., Krogfelt, K., and Wilcks, A. (2008). Selective pressure affects transfer and establishment of a *Lactobacillus plantarum* resistance plasmid in the gastrointestinal environment. J. Antimicrob. Chemother. *61*, 845–852.

Fey, P.D., and Olson, M.E. (2010). Current concepts in biofilm formation of *Staphylococcus epidermidis*. Future Microbiol. *5*, 917–913.

Fineran, P.C., Blower, T.R., Foulds, I.J., Humphreys, D.P., Lilley, K.S., and Salmond, G.P. (2009). The phage abortive infection system, ToxIN, functions as a protein-RNA toxin–antitoxin pair. Proc. Natl. Acad. Sci. U.S.A. *106*, 894–899.

Frigaard N.U., Martinez, A., Mincer, T.J., and DeLong, E.F. (2006). Proteorhodopsin lateral gene transfer between marine planktonic Bacteria and Archaea. Nature *439*, 847–850.

Frischer, M.E., Stewart, G.J., and Paul, J.H. (1994). Plasmid transfer to indigenous marine bacterial populations by natural transformation. FEMS Microbiol. Ecol. *15*, 127–135.

Frye, J.G., Lindsey, R.L., Meinersmann, R.J., Berrang, M.E., Jackson, C.R., Englen, M.D., Turpin, J.B., and Fedorka-Cray, P.J. (2011). Related antimicrobial resistance genes detected in different bacterial species co-isolated from swine fecal samples. Foodborne Pathog. Dis. *8*, 663–679.

Fuqua, W.C., and Winans, S.C. (1994). A LuxR-LuxI type regulatory system activates *Agrobacterium* Ti plasmid conjugal transfer in the presence of a plant tumor metabolite. J. Bacteriol. *176*, 2796–2806.

Fuqua, C., Winans, S.C., and Greenberg, E.P. (1996). Census and consensus in bacterial ecosystems: the LuxR-LuxI family of quorum-sensing transcriptional regulators. Annu. Rev. Microbiol. *50*, 727–751.

Fux, C.A., Costerton, J.W., Stewart, P.S., and Stoodley, P. (2005). Survival strategies of infectious biofilms. Trends Microbiol. *13*, 34–40.

Garbeva, P., van Veen, J.A., and van Elsas, J.D. (2004). Microbial diversity in soil: selection microbial populations by plant and soil type and implications for disease suppressiveness. Annu. Rev. Phytopathol. *42*, 243–270.

Garneau, J.E., Dupuis, M.È., Villion, M., Romero, D.A., Barrangou, R., Boyaval, P., Fremaux, C., Horvath, P., Magadán, A.H., and Moineau, S. (2010). The CRISPR/Cas bacterial immune system cleaves bacteriophage and plasmid DNA. Nature *468*, 67–71.

Ghigo, J.M. (2001). Natural conjugative plasmids induce bacterial biofilm development. Nature *412,* 442–445.

Ghosh, D., Roy, K., Williamson, K.E., White, D.C., Wommack, K.E., Sublette, K.L., and Radosevich, M. (2008). Prevalence of lysogeny among soil bacteria and presence of 16S rRNA and trzN genes in viral-community DNA. Appl. Environ. Microbiol. *74,* 495–502.

Ghosh, D., Roy, K., Williamson, K.E., Srinivasiah, S., Wommack, K.E., and Radosevich, M. (2009). Acyl-homoserine lactones can induce virus production in lysogenic bacteria: an alternative paradigm for prophage induction. Appl. Environ. Microbiol. *75,* 7142–7152.

Gibiansky, M.L., Conrad, J.C., Jin, F., Gordon, V.D., Motto, D.A., Mathewson, M.A., Stopka, W.G., Zelasko, D.C., Shrout, J.D., and Wong, G.C. (2010). Bacteria use type IV pili to walk upright and detach from surfaces. Science *330,* 197.

Gödeke, J., Paul, K., Lassak, J., and Thormann, K.M. (2011). Phage-induced lysis enhances biofilm formation in *Shewanella oneidensis* MR-1. ISME J. *5,* 613–626.

Goerke, C., and Wolz, C. (2010). Adaptation of *Staphylococcus aureus* to the cystic fibrosis lung. Int. J. Med. Microbiol. *300,* 520–525.

Goodman, A.E., Hild, E., Marshall, K.C., and Hermansson, M. (1993). Conjugative plasmid transfer between bacteria under simulated marine oligotrophic conditions. Appl. Environ. Microbiol. *59,* 1035–1040.

Götz, A and Smalla, K. (1997). Manure enhances plasmid mobilization and survival of *Pseudomonas putida* introduced into field soil. Appl. Environ. Microbiol. *63,* 1980–1986.

Götz, A., Pukall, R., Smit, E., Tietze, E., Prager, R., Tschäpe, H., van Elsas, J.D., and Smalla, K. (1996). Detection and characterization of broad-host-range plasmids in environmental bacteria by PCR. Appl. Environ. Microbiol. *62,* 2621–2628.

Graham, J.B., and Istock, C.A. (1978). Genetic exchange in *Bacillus subtilis* in soil. Mol. Gen. Genet. *116,* 287–298.

Greiner, L.L., Edwards, J.L., Shao, J., Rabinak, C., Entz, D., and Apicella, M.A. (2005). Biofilm formation by *Neisseria gonorrhoeae.* Infect. Immun. *73,* 1964–1970.

Griffith, F. (1928). The significance of pneumococcal types. J. Hyg. *27,* 113–159.

Guerin, E., Cambray, G., Sanchez-Alberola, N., Campoy, S., Erill, I., Da Re, S., Gonzalez-Zorn, B., Barbé, J., Ploy, M.C., and Mazel, D. (2009). The SOS response controls integron recombination. Science *324,* 1034.

Gulden, R.H., Lerat, S., Hart, M.M., Powell, J.R., Trevors, J.T., Pauls, K.P., Klironomos, J.N., and Swanton, C.J. (2005). Quantitation of transgenic plant DNA in leachate water: real-time polymerase chain reaction analysis. J. Agric. Food Chem. *53,* 5858–5865.

Gupta, P., Mathur, N., Bhatnagar, P., Nagar, P., and Srivastava, S. (2009). Genotoxicity evaluation of hospital wastewaters. Ecotoxicol Environ. Saf. *72,* 1925–1932.

Haddock, G., Mullin, M., MacCallum, A., Sherry, A., Tetley, L., Watson, E., Dagleish, M., Smith, D.G., and Everest, P. (2010). *Campylobacter jejuni* 81–176 forms distinct microcolonies on *in vitro*-infected human small intestinal tissue prior to biofilm formation. Microbiology *156,* 3079–3084.

Hader, D.P. (2000). Effects of solar UV-B radiation on aquatic ecosystems. Adv. Space Res. *26,* 2029–2040.

Haft, D.H., Selengut, J., Mongodin, E.F., and Nelson, K.E. (2005). A guild of 45 CRISPR-associated (Cas) protein families and multiple CRISPR/Cas subtypes exist in prokaryotic genomes. PLoS Comput. Biol. *1,* e60.

Hall-Stoodley, L., Costerton, J.W., and Stoodley, P. (2004). Bacterial biofilms: from the natural environment to infectious diseases. Nat. Rev. Microbiol. *2,* 95–108.

Hamilton-Brehm, S.D., Schut, G.J., and Adams, M.W. (2005). Metabolic and evolutionary relationships among *Pyrococcus* species: genetic exchange within a hydrothermal vent environment. J. Bacteriol. *187,* 7492–7499.

Hannan, S., Ready, D., Jasni, A.S., Rogers, M., Pratten, J., and Roberts, A.P. (2010). Transfer of antibiotic resistance by transformation with eDNA within oral biofilms. FEMS Immunol. Med. Microbiol. *59,* 345–349.

Harmsen, M., Yang, L., Pamp, S.J., and Tolker-Nielsen, T. (2010). An update on *Pseudomonas aeruginosa* biofilm formation, tolerance, and dispersal. FEMS Immunol. Med. Microbiol. *59,* 253–268.

Hassett, D.J., Korfhagen, T.R., Irvin, R.T., Schurr, M.J., Sauer, K., Lau, G.W., Sutton, M.D., Yu, H., and Hoiby, N. (2010). *Pseudomonas aeruginosa* biofilm infections in cystic fibrosis: insights into pathogenic processes and treatment strategies. Expert Opin. Ther. Targets *14,* 117–130.

Hastings, P.J., Rosenberg, S.M., and Slack, A. (2004). Antibiotic-induced lateral transfer of antibiotic resistance. Trends Microbiol. *12,* 401–404.

Hausner, M., and Wuertz, S. (1999). High rates of conjugation in bacterial biofilms as determined by quantitative *in situ* analysis. Appl. Environ. Microbiol. *65,* 3710–3713.

Häussler, S. (2010). Multicellular signalling and growth of *Pseudomonas aeruginosa.* Int. J. Med. Microbiol. *300,* 544–548.

Hayes, C.S., Aoki, S.K., and Low, D.A. (2010). Bacterial contact-dependent delivery systems. Annu. Rev. Genet. *44,* 71–90.

Hazen, T.H., Pan, L., Gu, J.D., and Sobecky, P.A. (2010). The contribution of mobile genetic elements to the evolution and ecology of Vibrios. FEMS Mircobiol. Ecol. *74,* 485–499.

He, X., Chang, W., Pierce, D.L., Seib, L.O., Wagner, J., and Fuqua, C. (2003). Quorum sensing in *Rhizobium* sp. strain NGR234 regulates conjugal transfer (tra) gene expression and influences growth rate. J. Bacteriol. *185,* 809–822.

He, Y.Y., Klisch, M., and Häder, D.P. (2002). Adaptation of cyanobacteria to UV-B stress correlated with oxidative stress and oxidative damage. Photochem. Photobiol. *76,* 188–196.

Heuer, H., and Smalla, K. (2007). Manure and sulfadiazine synergistically increased bacterial antibiotic resistance in soil over at least two months. Environ. Microbiol. *9,* 657–666.

Heuer, H., Kopmann, C., Binh, C.T., Top, E.M., and Smalla, K. (2009). Spreading antibiotic resistance through spread manure: characteristics of a novel plasmid type with low%G+C content. Environ. Microbiol. *11,* 937–949.

Hinnebusch, B.J., Rosso, M.L., Schwan, T.G., and Carniel, E. (2002). High-frequency conjugative transfer of antibiotic resistance genes to *Yersinia pestis* in the flea midgut. Mol. Microbiol. *46,* 349–354.

Hiom, K.J., and Sedgwick, S.G. (1992). Alleviation of EcoK DNA restriction in *Escherichia coli* and involvement of umuDC activity. Mol. Gen. Genet. *231,* 265–275.

Hoffman, L.R., D'Argenio, D.A., MacCoss, M.J., Zhang, Z., Jones, R.A., and Miller, S.I. (2005). Aminoglycoside antibiotics induce bacterial biofilm formation. Nature *436,* 1171–1175.

Høiby, N., Bjarnsholt, T., Givskov, M., Molin, S., and Ciofu, O. (2010a). Antibiotic resistance of bacterial biofilms. Int. J. Antimicrob. Agents *35,* 322–332.

Høiby, N., Ciofu, O., and Bjarnsholt, T. (2010b). *Pseudomonas aeruginosa* biofilms in cystic fibrosis. Future Microbiol. *5,* 1663–1667.

Hooper, S.D., Raes, J., Foerstner, K.U., Harrington, E.D., Dalevi, D., and Bork, P. (2008). A molecular study of microbe transfer between distant environments. PLoS One *3,* e2607.

Hughes, V.M., and Datta, N. (1983). Conjugative plasmids in bacteria of the 'pre-antibiotic' era. Nature *302,* 725–726.

Hyman, P., and Abedon, S.T. (2010). Bacteriophage host range and bacterial resistance. Adv. Appl. Microbiol. *70C,* 217–248.

Ilves, H., Horak, R., and Kivisaar, M. (2001). Involvement of sigma(S) in starvation-induced transposition of *Pseudomonas putida* transposon *Tn*4652. J. Bacteriol. *183,* 5445–5448.

Isidori, M., Lavorgna, M., Palumbo, M., Piccioli, V., and Parrella, A. (2007). Influence of alkylphenols and trace elements in toxic, genotoxic, and endocrine disruption activity of wastewater treatment plants. Environ. Toxicol Chem. *26,* 1686–1694.

Jaffe, A., Ogura, T., and Hiraga, S. (1985). Effects of the ccd function of the F plasmid on bacterial growth. J. Bacteriol. *163,* 841–849.

Jansen, R., Embden, J.D., Gaastra, W., and Schouls, L.M. (2002). Identification of genes that are associated with DNA repeats in prokaryotes. Mol. Microbiol. *43,* 1565–1575.

Jennings, J.C., Kolwyck, D.C., Kays, S.B., Whetsell, A.J., Surber, J.B., Cromwell, G.L., Lirette, R.P., and Glenn, K.C. (2003). Determining whether transgenic and endogenous plant DNA and transgenic protein are detectable in muscle from swine fed Roundup Ready soybean meal. J. Anim. Sci. *81,* 1447–1455.

Jeters, R.T., Wang, G.R., Moon, K., Shoemaker, N.B., and Salyers, A.A. (2009). Tetracycline-associated transcriptional regulation of transfer genes of the *Bacteroides* conjugative transposon CTnDOT. J. Bacteriol. *191,* 6374–6382.

Jiang, S.C., and Paul, J.H. (1998). Gene transfer by transduction in the marine environment. Appl. Environ. Microbiol. *64,* 2780–2787.

Johnsen, P.J., Simonsen, G.S., Olsvik, O., Midtvedt, T., and Sundsfjord, A. (2002). Stability, persistence, and evolution of plasmid-encoded VanA glycopeptide resistance in enterococci in the absence of antibiotic selection *in vitro* and in gnotobiotic mice. Microb. Drug Resist. *8,* 161–170.

Jolibois, B., and Guerbet, M. (2005). Evaluation of industrial, hospital and domestic wastewater genotoxicity with the *Salmonella* fluctuation test and the SOS chromotest. Mutat. Res. *565,* 151–162.

Jolibois, B., and Guerbet, M. (2006). Hospital wastewater genotoxicity. Ann. Occup. Hyg. *50,* 189–196.

Jolibois, B., Guerbet, M., and Vassal, S. (2003). Detection of hospital wastewater genotoxicity with the SOS chromotest and Ames fluctuation test. Chemosphere. *51*, 539–543.

Jones, B.V., Sun, F., and Marchesi, J.R. (2010). Comparative metagenomic analysis of plasmid encoded functions in the human gut microbiome. BMC Genomics *11*, 46.

Jones, S., Yu, B., Bainton, N.J., Birdsall, M., Bycroft, B.W., Chhabra, S.R., Cox, A.J.R., Golby, P., Reeves, P.J., Stephens, S., *et al.* (1993). The lux autoinducer regulates the production of exoenzyme virulence determinants in *Erwinia carotovora* and *Pseudomonas aeruginosa*. EMBO J. *12*, 2477–2482.

Juhas, M., van der Meer, J.R., Gaillard, M., Harding, R.M., Hood, D.W., and Crook, D.W. (2009). Genomic islands: tools of bacterial horizontal gene transfer and evolution. FEMS Mircobiol. Rev. *33*, 376–393.

Kaplan, J.B. (2010). Biofilm dispersal: mechanisms, clinical implications, and potential therapeutic uses. J. Dent. Res. *89*, 205–218.

Karavolos, M.H., Spencer, H., Bulmer, D.M., Thompson, A., Winzer, K., Williams, P., Hinton, J.C., and Khan, C.M. (2008). Adrenaline modulates the global transcriptional profile of *Salmonella* revealing a role in the antimicrobial peptide and oxidative stress resistance responses. BMC Genomics *9*, 458.

Kay, M.K., Erwin, T.C., McLean, R.J., and Aron, G.M. (2011). Bacteriophage ecology in *Escherichia coli* and *Pseudomonas aeruginosa* mixed-biofilm communities. Appl. Environ. Microbiol. *77*, 821–829.

Kelleher, J.E., and Raleigh, E.A. (1994). Response to UV damage by four *Escherichia coli* K-12 restriction systems. J. Bacteriol. *176*, 5888–5896.

Kenzaka, T., Tani, K., and Nasu, M. (2010). High-frequency phage-mediated gene transfer in freshwater environments determined at single-cell level. ISME J. *4*, 648–659.

Khachatryan, Z.A., Ktsoyan, Z.A., Manukyan, G.P., Kelly, D., Ghazaryan, K.A., and Aminov, R.I. (2008). Predominant role of host genetics in controlling the composition of gut microbiota. PLoS ONE 3, e3064.

Koonin, E.V., Makarova, K.S., and Aravind, L. (2001). Horizontal gene transfer in prokaryotes: quantification and classification. Annu. Rev. Microbiol. *55*, 709–742.

Kroer, N., Barkay, T., Sörensen, S., and Weber, D. (1998). Effect of root exudates and bacterial metabolic activity on conjugative gene transfer in the rhizosphere of a marsh plant. FEMS Microbiol. Ecol. *25*, 375–384.

Kurokawa, K., Itoh, T., Kuwahara, T., Oshima, K., Toh, H., Toyoda, A., Takami, H., Morita, H., Sharma, V.K., Srivastava, T.P., *et al.* (2007). Comparative metagenomics revealed commonly enriched gene sets in human gut microbiomes. DNA Res. *14*, 169–181.

Kusumoto, M., Ooka, T., Nishiya, Y., Ogura, Y., Saito, T., Sekine, Y., Iwata, T., Akiba, M., and Hayashi, T. (2011). Insertion sequence-excision enhancer removes transposable elements from bacterial genomes and induces various genomic deletions. Nat. Commun. *2*, 152.

Labrie, S.J., Samson, J.E., and Moineau, S. (2010). Bacteriophage resistance mechanisms. Nat. Rev. Microbiol. *8*, 317–327.

Lang, A.S., and Beatty, J.T. (2007). Importance of widespread gene transfer agent genes in alpha-proteobacteria. Trends Microbiol. *15*, 54–62.

Lederberg, J. (1952). Cell genetics and hereditary symbiosis. Physiol. Rev. *32*, 403–430.

Lerat, S., Gulden, R.H., Hart, M.M., Powell, J.R., England, L.S., Pauls, K.P., Swanton, C.J., Klironomos, J.N., and Trevors, J.T. (2007). Quantification and persistence of recombinant DNA of Roundup Ready corn and soybean in rotation. J. Agric. Food Chem. *55*, 10226–10231.

Levy, M.S., Balbinder, E., and Nagel, R. (1993). Effect of mutations in SOS genes on UV-induced precise excision of *Tn*10 in *Escherichia coli*. Mutat. Res. *293*, 241–247.

Levy-Booth, D.J., Campbell, R.G., Gulden, R.H., Hart, M.M., Powell, J.R., Klironomos, J.N., Pauls, K.P., Swanton, C.J., Trevors, J.T., and Dunfield, K.E. (2008). Real-time polymerase chain reaction monitoring of recombinant DNA entry into soil from decomposing roundup ready leaf biomass. J. Agric. Food Chem. *56*, 6339–6347.

Li, L.-Y., Shoemaker, N.B., and Salyers, A.A. (1993). Characterization of the mobilization region of a *Bacteroides* insertion element (NBU1) that is excised and transferred by *Bacteroides* conjugative transposons. J. Bacteriol. *175*, 6588–6598.

Li, Y., Hao, G., Galvani, C.D., Meng, Y., De La Fuente, L., Hoch, H.C., and Burr, T.J. (2007). Type I and type IV pili of *Xylella fastidiosa* affect twitching motility, biofilm formation and cell–cell aggregation. Microbiology *153*, 719–726.

Liebert, C.A., Hall, R.M., and Summers, A.O. (1999). Transposon *Tn*21, flagship of the floating genome. Mircobiol. Mol. Biol. Rev. *63*, 507–522.

Lilley, A.K., and Bailey, M.J. (1997). The acquisition of indigenous plasmids by a genetically marked pseudomonad population colonizing the sugar beet phytosphere is related to local environmental conditions. Appl. Environ. Microbiol. *63,* 1577–1583.

Lilley, A.K., Fry, J.C., Day, M.J., and Bailey, M.J. (1994). In situ transfer of an exogenously isolated plasmid between indigenous donor and recipient *Pseudomonas* spp. in sugar beet rhizosphere. Microbiology *140,* 27–33.

Lindsay, D., and von Holy, A. (2006). Bacterial biofilms within the clinical setting: what healthcare professionals should know. J. Hosp. Infect. *64,* 313–325.

Little, J.W. (2005). Lysogeny, prophage induction, and lysogenic conversion, In Phages, Waldor, M.K., Friedman, D.I., and Adhya, S., eds (ASM Press, Washington, DC), pp. 37–54.

López-García, P., Brochier, C., Moreira, D., and Rodríguez-Valera, F. (2004). Comparative analysis of a genome fragment of an uncultivated mesopelagic crenarchaeote reveals multiple horizontal gene transfers. Environ. Microbiol. *6,* 19–34.

Lorenz, M.G., Aardema, B.W., and Wackernagel, W. (1988). Highly efficient genetic transformation of *Bacillus subtilis* attached to sand grains. J. Gen. Microbiol. *134,* 107–112.

Lorenz, M.G., and Wackernagel, W. (1990). Natural genetic transformation of *Pseudomonas stutzeri* by sand-adsorbed DNA. Arch. Microbiol. *154,* 380–385.

Louis, P., McCrae, S.I., Charrier, C., and Flint, H.J. (2007). Organization of butyrate synthetic genes in human colonic bacteria: phylogenetic conservation and horizontal gene transfer. FEMS Microbiol. Lett. *269,* 240–247.

Luke, N.R., Jurcisek, J.A., Bakaletz, L.O., and Campagnari, A.A. (2007). Contribution of *Moraxella catarrhalis* type IV pili to nasopharyngeal colonization and biofilm formation. Infect. Immun. *75,* 5559–5564.

McDaniel, L.D., Young, E., Delaney, J., Ruhnau, F., Ritchie, K.B., and Paul, J.H. (2010). High frequency of horizontal gene transfer in the oceans. Science *330,* 50.

McKenzie, R.L., Aucamp, P.J., Bais, A.F., Björn, L.O., and Ilyas, M. (2007). Changes in biologically-active ultraviolet radiation reaching the Earth's surface. Photochem. Photobiol. Sci. *6,* 218–231.

Maeda, S., Ito, M., Ando, T., Ishimoto, Y., Fujisawa, Y., Takahashi, H., Matsuda, A., Sawamura, A., and Kato, S. (2006). Horizontal transfer of nonconjugative plasmids in a colony biofilm of *Escherichia coli*. FEMS Microbiol. Lett. *255,* 115–120.

Mahillon J., and Chandler, M. (1998). Insertion sequences. Mircobiol. Mol. Biol. Rev. *62,* 725–774.

Maiques, E., Ubeda, C., Campoy, S., Salvador, N., Lasa, I., Novick, R.P., Barbé, J., and Penadés, J.R. (2006). β-lactam antibiotics induce the SOS response and horizontal transfer of virulence factors in *Staphylococcus aureus*. J. Bacteriol. *188,* 2726–2729.

Matsuo, J., Oguri, S., Nakamura, S., Hanawa, T., Fukumoto, T., Hayashi, Y., Kawaguchi, K., Mizutani, Y., Yao, T., Akizawa, K., *et al.* (2010). Ciliates rapidly enhance the frequency of conjugation between *Escherichia coli* strains through bacterial accumulation in vesicles. Res. Microbiol. *161,* 711–719.

Mazel, D. (2006). Integrons: agents of bacterial evolution. Nat. Rev. Microbiol. *4,* 608–620.

Meibom, K.L., Blokesch, M., Dolganov, N.A., Wu, C.Y., and Schoolnik, G.K. (2005). Chitin induces natural competence in *Vibrio cholerae*. Science *310,* 1824–1827.

Melechen, N.E., and Go, G. (1980). Induction of lambdoid prophages by amino acid deprivation: differential inducibility; role of recA. Mol. Gen. Genet. *180,* 147–155.

Miller, C., Thomsen, L.E., Gaggero, C., Mosseri, R., Ingmer, H., and Cohen, S.N. (2004). SOS response induction by beta-lactams and bacterial defense against antibiotic lethality. Science. *305,* 1629–1631.

Mohr, K.I., and Tebbe, C.C. (2007). Field study results on the probability and risk of a horizontal gene transfer from transgenic herbicide-resistant oilseed rape pollen to gut bacteria of bees. Appl. Mircobiol. Biotechnol. *75,* 573–582.

Mølbak, L., Molin, S., and Kroer, N. (2007). Root growth and exudate production define the frequency of horizontal plasmid transfer in the Rhizosphere. FEMS Mircobiol. Ecol. *59,* 167–176.

Molin, S., and Tolker-Nielsen, T. (2003). Gene transfer occurs with enhanced efficiency in biofilms and induces enhanced stabilisation of the biofilm structure. Curr. Opin. Biotechnol. *14,* 255–261.

Moon, K., Shoemaker, N.B., Gardner, J.F., and Salyers, A.A. (2005). Regulation of excision genes of the *Bacteroides* conjugative transposon CTnDOT. J. Bacteriol. *187,* 5732–5741.

Moore, I.F., Hughes, D.W., and Wright, G.D. (2005). Tigecycline is modified by the flavin-dependent monooxygenase TetX. Biochemistry *44,* 11829–11835.

Morales, D.K., and Hogan, D.A. (2010). *Candida albicans* interactions with bacteria in the context of human health and disease. PLoS Pathog. *6,* e1000886.

Morrison, W.D., Miller, R.V., and Sayler, G.S. (1978). Frequency of F116-mediated transduction of *Pseudomonas aeruginosa* in a freshwater environment. Appl. Environ. Microbiol. *36,* 724–730.

Moscoso, M., García, E., and López, R. (2009). Pneumococcal biofilms. Int. Microbiol. *12,* 77–85.

Musovic, S., Oregaard, G., Kroer, N., and Sørensen, S.J. (2006). Cultivation-independent examination of horizontal transfer and host range of an IncP-1 plasmid among Gram-positive and Gram-negative bacteria indigenous to the barley rhizosphere. Appl. Environ. Microbiol. *72,* 6687–6692.

Naito, M., Frirdich, E., Fields, J.A., Pryjma, M., Li, J., Cameron, A., Gilbert, M., Thompson, S.A., and Gaynor, E.C. (2010). Effects of sequential *Campylobacter jejuni* 81–176 lipooligosaccharide core truncations on biofilm formation, stress survival, and pathogenesis. J. Bacteriol. *192,* 2182–2192.

Netherwood T., Martín-Orúe, S.M., O'Donnell, A.G., Gockling, S., Graham, J., Mathers, J.C., and Gilbert, H.J. (2004). Assessing the survival of transgenic plant DNA in the human gastrointestinal tract. Nat. Biotechnol. *22,* 204–209.

Nielsen, K.M., Bones, A.M., Smalla, K., and van Elsas, J.D. (1998). Horizontal gene transfer from transgenic plants to terrestrial bacteria – a rare event? FEMS Mircobiol. Rev. *22,* 79–103.

Novick, R.P. (1987). Plasmid incompatibility. Microbiol. Rev. *51,* 381–395.

Ochman, H., Lawrence, J.G., and Groisman, E.A. (2000). Lateral gene transfer and the nature of bacterial innovation. Nature *405,* 299–304.

Ogura, T., and Hiraga, S. (1983). Mini-F plasmid genes that couple host cell division to plasmid proliferation. Proc. Natl. Acad. Sci. U.S.A. *80,* 4784–4788.

Oguri, S., Matsuo, J., Hayashi, Y., Nakamura, S., Hanawa, T., Fukumoto, T., Mizutani, Y., Yao, T., Akizawa, K., Suzuki, H., *et al.* (2011). Ciliates promote the transfer of the gene encoding the extended-spectrum beta-lactamase CTX-M-27 between *Escherichia coli* strains. J. Antimicrob. Chemother. *66,* 527–530.

Ooka, T., Ogura, Y., Asadulghani, M., Ohnishi, M., Nakayama, K., Terajima, J., Watanabe, H., and Hayashi, T. (2009). Inference of the impact of insertion sequence (IS) elements on bacterial genome diversification through analysis of small-size structural polymorphisms in *Escherichia coli* O157 genomes. Genome Res. *19,* 1809–1816.

Palenik, B., Ren, Q., Tai, V., and Paulsen, I.T. (2009). Coastal *Synechococcus* metagenome reveals major roles for horizontal gene transfer and plasmids in population diversity. Environ. Microbiol. *11,* 349–359.

Parsek, M.R., and Singh, P.K. (2003). Bacterial biofilms: an emerging link to disease pathogenesis. Annu. Rev. Microbiol. *57,* 677–701.

Passador, L., Cook, J.M., Gambello, M.J., Rust, L., and Iglewski, B.H. (1993). Expression of *Pseudomonas aeruginosa* virulence genes requires cell-to-cell communication. Science *260,* 1127–1130.

Paul, J.H., Frischer, M.E., and Thurmond, J.M. (1991). Gene transfer in marine water column and sediment microcosms by natural plasmid transformation. Appl. Environ. Microbiol. *57,* 1509–1515.

Paul, J.H. (2008). Prophages in marine bacteria: dangerous molecular time bombs or the key to survival in the seas? ISME J. *2,* 579–589.

Peleg, A.Y., Adams, J., and Paterson, D.L. (2007). Tigecycline efflux as a mechanism for nonsusceptibility in *Acinetobacter baumannii.* Antimicrob. Agents Chemother. *51,* 2065–2069.

Petersen, J., Brinkmann, H., and Pradella, S. (2009). Diversity and evolution of *repABC* type plasmids in Rhodobacterales. Environ. Microbiol. *11,* 2627–2638.

Petersen, J., Brinkmann, H., Berger, M., Brinkhoff, T., Päuker, O., and Pradella, S. (2011). Origin and evolution of a novel DnaA-like plasmid replication type in Rhodobacterales. Mol. Biol. Evol. *28,* 1229–1240.

Peterson, G., Kumar, A., Gart, E., and Narayanan, S. (2011). Catecholamines increase conjugative gene transfer between enteric bacteria. Microb. Pathog. *51,* 1–8.

Petridis, M., Bagdasarian, M., Waldor, M.K., and Walker, E. (2006). Horizontal transfer of Shiga toxin and antibiotic resistance genes among *Escherichia coli* strains in house fly (Diptera: Muscidae) gut. J. Med. Entomol. *43,* 288–295.

Phipps, R.H., Jones, A.K., Tingey, A.P., and Abeyasekera, S. (2005). Effect of corn silage from an herbicide-tolerant genetically modified variety on milk production and absence of transgenic DNA in milk. J. Dairy Sci. *88,* 2870–2878.

Pierson, L.S., III, Keppenne, V.D., and Wood, D.W. (1994). Phenazine antibiotic biosynthesis in *Pseudomonas aureofaciens* 30–84 is regulated by phzR in response to cell density. J. Bacteriol. *176,* 3966–3974.

Piper, K.R., Beck von Bodman, S., and Farrand, S.K. (1993). Conjugation factor of *Agrobacterium tumefaciens* regulates Ti plasmid transfer by autoinduction. Nature *362,* 448–450.

Pirhonen, M., Flego, D., Heikinheimo, R., and Palva, E.T. (1993). A small diffusible signal molecule is responsible for the global control of virulence and exoenzyme production in the plant pathogen, *Erwinia carotovora*. EMBO J. *12,* 2467–2476.

Pontiroli, A., Rizzi, A., Simonet, P., Daffonchio, D., Vogel, T.M., and Monier, J.M. (2009). Visual evidence of horizontal gene transfer between plants and bacteria in the phytosphere of transplastomic tobacco. Appl. Environ. Microbiol. *75,* 3314–3322.

Poole, T and Crippen, T. (2009). Conjugative plasmid transfer between *Salmonella enterica* Newport and *Escherichia coli* within the gastrointestinal tract of the lesser mealworm beetle, *Alphitobius diaperinus* (Coleoptera: Tenebrionidae). Poult. Sci. *88,* 1553–1558.

Pradella, S., Päuker, O., and Petersen, J. (2010). Genome organization of the marine *Roseobacter* clade member *Marinovum algicola*. Arch. Microbiol. *192,* 115–126.

Prudhomme, M., Attaiech, L., Sanchez, G., Martin, B., and Claverys, J.P. (2006). Antibiotic stress induces genetic transformability in the human pathogen *Streptococcus pneumoniae*. Science *313,* 89–92.

Pukall, R., Tschape, H., and Smalla, K. (1996). Monitoring the spread of broad host and narrow host range plasmids in soil microcosms. FEMS Microbiol. Ecol. *20,* 53–66.

Pul, U., Wurm, R., Arslan, Z., Geissen, R., Hofmann, N., and Wagner, R. (2010). Identification and characterization of *E. coli* CRISPR-cas promoters and their silencing by H-NS. Mol. Mircobiol. *75,* 1495–1512.

Qin, Y., Keenan, C., and Farrand, S.K. (2009). N- and C-terminal regions of the quorum-sensing activator TraR cooperate in interactions with the alpha and sigma-70 components of RNA polymerase. Mol. Microbiol. *74,* 330–346.

Ramsay, J.P., Sullivan, J.T., Jambari, N., Ortori, C.A., Heeb, S., Williams, P., Barrett, D.A., Lamont, I.L., and Ronson, C.W. (2009). A LuxRI-family regulatory system controls excision and transfer of the *Mesorhizobium loti* strain R7A symbiosis island by activating expression of two conserved hypothetical genes. Mol. Microbiol. *73,* 1141–1155.

van Randen J., and Venema, G. (1984). Direct plasmid transfer from replica-plated *E. coli* colonies to competent *B. subtilis* cells. Identification of an *E. coli* clone carrying the *hisH* and *tyrA* genes of *B. subtilis*. Mol. Gen. Genet. *195,* 57–61.

Rastogi, R.P., Richa Kumar, A., Tyagi, M.B., and Sinha, R.P. (2010). Molecular mechanisms of ultraviolet radiation-induced DNA damage and repair. J. Nucleic Acids *2010,* 592980.

Ravatn, R., Zehnder, A.J., and van der Meer, J.R. (1998). Low-frequency horizontal transfer of an element containing the chlorocatechol degradation genes from *Pseudomonas* sp. strain B13 to *Pseudomonas putida* F1 and to indigenous bacteria in laboratory-scale activated-sludge microcosms. Appl. Environ. Microbiol. *64,* 2126–2132.

Ray, J.L., Andersen, H.K., Young, S., Nielsen, K.M., and O'Callaghan, M. (2007). An assessment of the potential of herbivorous insect gut bacteria to develop competence for natural transformation. Environ. Biosafety Res. *6,* 135–147.

Reisner, A., Höller, B.M., Molin, S., and Zechner, E.L. (2006). Synergistic effects in mixed *Escherichia coli* biofilms: conjugative plasmid transfer drives biofilm expansion. J. Bacteriol. *188,* 3582–3588.

Reyes, A., Haynes, M., Hanson, N., Angly, F.E., Heath, A.C., Rohwer, F., and Gordon, J.I. (2010). Viruses in the faecal microbiota of monozygotic twins and their mothers. Nature *466,* 334–338.

Richaume, A., Angle, J.S., and Sadowsky, M.J. (1989). Influence of soil variables on in situ plasmid transfer from *Escherichia coli* to *Rhizobium fredii*. Appl. Environ. Microbiol. *55,* 1730–1734.

Richaume, A., Smit, E., Faurie, G., and van Elsas, J.D. (1992). Influence of soil type on the transfer of plasmid RP4p from *Pseudomonas fluorescens* to introduced recipient and to indigenous bacteria. FEMS Microbiol. Ecol. *101,* 281–292.

Richter, B., and Smalla, K. (2007). Screening of rhizosphere and soil bacteria for transformability. Environ. Biosafety Res. *6,* 91–99.

Roberts, A.P., and Mullany, P. (2009). A modular master on the move: the Tn*916* family of mobile genetic elements. Trends Microbiol. *17,* 251–258.

Rochelle, P.A., Fry, J.C., and Day, M.J. (1989). Factors affecting conjugal transfer of plasmids encoding mercury resistance from pure cultures and mixed natural suspensions of epilithic bacteria. J. Gen. Microbiol. *135,* 409–424.

Rohde, H., Frankenberger, S., Zähringer, U., and Mack, D. (2010). Structure, function and contribution of polysaccharide intercellular adhesin (PIA) to *Staphylococcus epidermidis* biofilm formation and pathogenesis of biomaterial-associated infections. Eur. J. Cell Biol. *89,* 103–111.

Romanowski, G., Lorenz, M.G., and Wackernagel, W. (1991). Adsorption of plasmid DNA to mineral surfaces and protection against DNase I. Appl. Environ. Microbiol. *57,* 1057–1061.

Romero, D., Pérez-García, A., Veening, J.W., de Vicente, A., and Kuipers, O.P. (2006). Transformation of undomesticated strains of *Bacillus subtilis* by protoplast electroporation. J. Mircobiol. Methods *66,* 556–559.

Salyers, A.A. (1993). Gene transfer in the mammalian intestinal tract. Curr. Opin. Biotechnol. *4,* 294–298.

Salyers, A.A., Gupta, A., and Wang, Y. (2004). Human intestinal bacteria as reservoirs for antibiotic resistance genes. Trends Microbiol. *12,* 412–416.

Schwartz, T., Jungfer, C., Heissler, S., Friedrich, F., Faubel, W., and Obst, U. (2009). Combined use of molecular biology taxonomy, Raman spectrometry, and ESEM imaging to study natural biofilms grown on filter materials at waterworks. Chemosphere *77,* 249–257.

Sengeløv, G., Kowalchuk, G.A., and Sørensen, S.J. (2000). Influence of fungal–bacterial interactions on bacterial conjugation in the residuesphere. FEMS Microbiol. Ecol. *31,* 39–45.

Sentchilo, V., Ravatn, R., Werlen, C., Zehnder, A.J., and van der Meer, J.R. (2003). Unusual integrase gene expression on the clc genomic island in *Pseudomonas* sp. strain B13. J. Bacteriol. *185,* 4530–4538.

Sharon, I., Alperovitch, A., Rohwer, F., Haynes, M., Glaser, F., Atamna-Ismaeel, N., Pinter, R.Y., Partensky, F., Koonin, E.V., Wolf, Y.I., *et al.* (2009). Photosystem I gene cassettes are present in marine virus genomes. Nature *461,* 258–262.

Sharon, I., Battchikova, N., Aro, E.M., Giglione, C., Meinnel, T., Glaser, F., Pinter, R.Y., Breitbart, M., Rohwer, F., and Béjà, O. (2011). Comparative metagenomics of microbial traits within oceanic viral communities. ISME J. *5,* 1178–1190.

Shime-Hattori, A., Iida, T., Arita, M., Park, K.S., Kodama, T., and Honda, T. (2006). Two type IV pili of *Vibrio parahaemolyticus* play different roles in biofilm formation. FEMS Microbiol. Lett. *264,* 89–97.

Shiner, E.K., Rumbaugh, K.P., and Williams, S.C. (2005). Inter-kingdom signaling: deciphering the language of acyl homoserine lactones. FEMS Mircobiol. Rev. *29,* 935–947.

Shoemaker, N.B., Wang, G.-R., Stevens, A.M., and Salyers, A.A. (1993). Excision, transfer, and integration of NBU1, a mobilizable site-selective insertion element. J. Bacteriol. *175,* 6578–6587.

Shoemaker, N.B., Wang, G.R., and Salyers, A.A. (2000). Multiple gene products and sequences required for excision of the mobilizable integrated *Bacteroides* element NBU1. J. Bacteriol. *182,* 928–936.

Showsh, S.A., and Andrews, R.E., Jr. (1992). Tetracycline enhances Tn*916*-mediated conjugal transfer. Plasmid *28,* 213–224.

Senadheera, D., and Cvitkovitch, D.G. (2008). Quorum sensing and biofilm formation by *Streptococcus mutans.* Adv. Exp. Med. Biol. *631,* 178–188.

Silverman, R.J., Nobbs, A.H., Vickerman, M.M., Barbour, M.E., and Jenkinson, H.F. (2010). Interaction of *Candida albicans* cell wall Als3 protein with *Streptococcus gordonii* SspB adhesin promotes development of mixed-species communities. Infect. Immun. *78,* 4644–4652.

Sinha, R.P., and Hader, D.P. (2002). Life under solar UV radiation in aquatic organisms. Adv. Space Res. *30,* 1547–1556.

Sizemore, R.K., and Colwell, R.R. (1977). Plasmids carried by antibiotic-resistant marine bacteria. Antimicrob. Agents Chemother. *12,* 373–382.

Smalla, K., Krögerrecklenfort, E., Heuer, H., Dejonghe, W., Top, E., Osborn, M., Niewint, J., Tebbe, C., Barr, M., Bailey, M., *et al.* (2000). PCR-based detection of mobile genetic elements in total community DNA. Microbiology *146,* 1256–1257.

Smiles, D.E. (1988). Aspects of the physical environment of soil organisms. Biol. Fertil. Soils *6,* 204–215.

Smillie, C., Garcillán-Barcia, M.P., Francia, M.V., Rocha, E.P., and de la Cruz, F. (2010). Mobility of plasmids. Mircobiol. Mol. Biol Rev. *74,* 434–452.

Smith, D.H. (1967). R factors mediate resistance to mercury, nickel and cobalt. Science *156,* 1114–1116.

Sobecky, P.A., and Coombs, J.M. (2009). Horizontal gene transfer in metal and radionuclide contaminated soils. Methods Mol. Biol. *532,* 455–472.

Sørensen, S.J., Schyberg, T., and Rønn, R. (1999). Predation by protozoa on *Escherichia coli* K12 in soil and transfer of resistance plasmid RP4 to indigenous bacteria in soil. Appl. Soil Ecol. *11,* 79–90.

Sørensen, S.J., Bailey, M., Hansen, L.H., Kroer, N., and Wuertz, S. (2005). Studying plasmid horizontal transfer *in situ*: a critical review. Nat. Rev. Microbiol. *3,* 700–710.

Spencer, H., Karavolos, M.H., Bulmer, D.M., Aldridge, P., Chhabra, S.R., Winzer, K., Williams, P., and Khan, C.M. (2010). Genome-wide transposon mutagenesis identifies a role for host neuroendocrine stress hormones in regulating the expression of virulence genes in *Salmonella.* J. Bacteriol. *192,* 714–724.

Sperandio, V., Torres, A.G., Jarvis, B., Nataro, J.P., and Kaper, J.B. (2003). Bacteria-host communication: the language of hormones. Proc. Natl. Acad. Sci. U.S.A. *100,* 8951–8956.

Stanton, T.B. (2007). Prophage-like gene transfer agents-novel mechanisms of gene exchange for *Methanococcus, Desulfovibrio, Brachyspira,* and *Rhodobacter* species. Anaerobe *13,* 43–49.

Stanton, T.B., Humphrey, S.B., Sharma, V.K., and Zuerner, R.L. (2008). Collateral effects of antibiotics: carbadox and metronidazole induce VSH-1 and facilitate gene transfer among *Brachyspira hyodysenteriae* strains. Appl. Environ. Microbiol. *74,* 2950–2956.

Steichen, C.T., Cho, C., Shao, J.Q., and Apicella, M.A. (2011). The *Neisseria gonorrhoeae* biofilm matrix contains DNA, and an indogenous nuclease controls its incorporation. Infect. Immun. *79,* 1504–1511.

Stepanauskas, R., Glenn, T.C., Jagoe, C.H., Tuckfield, R.C., Lindell, A.H., and McArthur, J.V. (2005). Elevated microbial tolerance to metals and antibiotics in metal-contaminated industrial environments. Environ. Sci. Technol. *39,* 3671–3678.

Stepanauskas, R., Glenn, T.C., Jagoe, C.H., Tuckfield, R.C., Lindell, A.H., King, C.J., and McArthur, J.V. (2006). Coselection for microbial resistance to metals and antibiotics in freshwater microcosms. Environ. Microbiol. *8,* 1510–1514.

Stephens, P.M., O'Sullivan, M., and O'Gara, F. (1987). Effect of bacteriophage on colonization of sugarbeet roots by fluorescent *Pseudomonas* spp. Appl. Environ. Microbiol. *53,* 1164–1167.

Stevens, A.M., Shoemaker, N.B., Li, L.Y., and Salyers, A.A. (1993). Tetracycline regulation of genes on *Bacteroides* conjugative transposons. J. Bacteriol. *175,* 6134–6141.

Stewart, G.J., and Sinigalliano, C.D. (1990). Detection of horizontal gene transfer by natural transformation in native and introduced species of bacteria in marine and synthetic sediments. Appl. Environ. Microbiol. *56,* 1818–1824.

Stewart, P.S., and Costerton, J.W. (2001). Antibiotic resistance of bacteria in biofilms. Lancet *358,* 135–138.

Stoebel, D.M., Free, A., and Dorman, C.J. (2008). Anti-silencing: overcoming H-NS-mediated repression of transcription in Gram-negative enteric bacteria. Microbiology *154,* 2533–2545.

Su, Y.A., He, P., and Clewell, D.B. (1992). Characterization of the *tet*(M) determinant of Tn*916*: evidence for regulation by transcription attenuation. Antimicrob. Agents Chemother. *36,* 769–778.

Teh, K.H., Flint, S., and French, N. (2010). Biofilm formation by *Campylobacter jejuni* in controlled mixed-microbial populations. Int. J. Food Microbiol. *143,* 118–124.

Thimm, T., Hoffmann, A., Fritz, I., and Tebbe, C.C. (2001). Contribution of the earthworm *Lumbricus rubellus* (Annelida, Oligochaeta) to the establishment of plasmids in soil bacterial communities. Microb. Ecol. *41,* 341–351.

Timms-Wilson, T.M., Van Overbeek, L.S., Bailey, M.J., Trevors, J.T., and Van Elsas, J.D. (2001). Quantification of gene transfer in soil and the rhizosphere. In Manual of Environmental Microbiology, Hurst, C.J., Crawford, R.L., Knudsen, G.R., McInerney, M.J., and Stetzenbach, L.D., eds (ASM Press, Washington, DC), pp. 648–659.

Toba, F.A., Thompson, M.G., Campbell, B.R., Junker, L.M., Ruggeberg, K.G., and Hay, A.G. (2011). Role of DLP12 lysis genes in *E. coli* biofilm formation. Microbiology *157,* 1640–1650.

Toleman, M.A., and Walsh, T.R. (2010). ISCR elements are key players in IncA/C plasmid evolution. Antimicrob. Agents Chemother. *54,* 3534.

Toleman, M.A., Bennett, P.M., and Walsh, T.R. (2006). ISCR elements: novel gene-capturing systems of the 21st century? Mircobiol. Mol. Biol. Rev. *70,* 296–316.

Toomey, N., Monaghan, A., Fanning, S., and Bolton, D. (2009). Transfer of antibiotic resistance marker genes between lactic acid bacteria in model rumen and plant environments. Appl. Environ. Microbiol. *75,* 3146–3152.

Top, E.M., De Rore, H., Collard, J.M., Gellens, V., and Slobodkina, G. (1995). Retromobilization of heavy metal resistance genes in unpolluted and heavy metal polluted soil. FEMS Microbiol. Ecol. *18,* 191–203.

Torres, O.R., Korman, R.Z., Zahler, S.A., and Dunny, G.M. (1991). The conjugative transposon Tn*925*: enhancement of conjugal transfer by tetracycline in *Enterococcus faecalis* and mobilization of chromosomal genes in *Bacillus subtilis* and *E. faecalis.* Mol. Gen. Genet. *225,* 395–400.

Trevors, J.T., Van Elsas, J.D., Van Overbeek, L.S., and Starodub, M.E. (1990). Transport of a genetically engineered *Pseudomonas fluorescens* strain through a soil microcosm. Appl. Environ. Microbiol. *56,* 401–408.

Tun-Garrido, C., Bustos, P., González, V., and Brom, S. (2003). Conjugative transfer of p42a from *Rhizobium etli* CFN42, which is required for mobilization of the symbiotic plasmid, is regulated by quorum sensing. J. Bacteriol. *185,* 1681–1692.

Ubeda, C., Maiques, E., Knecht, E., Lasa, I., Novick, R.P., and Penadés, J.R. (2006). Antibiotic-induced SOS response promotes horizontal dissemination of pathogenicity island-encoded virulence factors in staphylococci. Mol. Microbiol. *56,* 836–844.

Valentine, P.J., Shoemaker, N.B., and Salyers, A.A. (1988). Mobilization of *Bacteroides* plasmids by *Bacteroides* conjugal elements. J. Bacteriol. *170,* 1319–1324.

Van Melderen, L. (2010). Toxin–antitoxin systems: why so many, what for? Curr. Opin. Microbiol. *13*, 781–785.
Varga, J.J., Therit, B., and Melville, S.B. (2008). Type IV pili and the CcpA protein are needed for maximal biofilm formation by the Gram-positive anaerobic pathogen *Clostridium perfringens*. Infect. Immun. *76*, 4944–4951.
Vos, M., Birkett, P.J., Birch, E., Griffiths, R.I., and Buckling, A. (2009). Local adaptation of bacteriophages to their bacterial hosts in soil. Science *325*, 833.
Wang, X., Li, M., Yan, Q., Chen, X., Geng, J., Xie, Z., and Shen, P. (2007). Across genus plasmid transformation between *Bacillus subtilis* and *Escherichia coli* and the effect of *Escherichia coli* on the transforming ability of free plasmid DNA. Curr. Microbiol. *54*, 450–456.
Wang, X., Kim, Y., and Wood, T.K. (2009). Control and benefits of CP4–57 prophage excision in *Escherichia coli* biofilms. ISME J. *3*, 1164–1179.
Wang, X., Kim, Y., Ma, Q., Hong, S.H., Pokusaeva, K., Sturino, J.M., and Wood, T.K. (2010). Cryptic prophages help bacteria cope with adverse environments. Nat. Commun. *1*, 147.
Wang, Y., Rotman, E.R., Shoemaker, N.B., and Salyers, A.A. (2005). Translational control of tetracycline resistance and conjugation in the *Bacteroides* conjugative transposon CTnDOT. J. Bacteriol. *187*, 2673–2680.
Watanabe, T. (1963). Infective heredity of multiple drug resistance in bacteria. Bacteriol. Rev. *27*, 87–115.
Watanabe, T., and Fukasawa, T. (1961). Episome-mediated transfer of drug resistance in Enterobacteriaceae. I. Transfer of resistance factors by conjugation. J. Bacteriol. *81*, 669–678.
Weinberg, S.R., and Stotzky, G. (1972). Conjugation and genetic recombination of *Escherichia coli* in soil. Soil Biol. Biochem. *4*, 171–180.
Westra, E.R., Pul, U., Heidrich, N., Jore, M.M., Lundgren, M., Stratmann, T., Wurm, R., Raine, A., Mescher, M., Van Heereveld, L., *et al.* (2010). H-NS-mediated repression of CRISPR-based immunity in *Escherichia coli* K12 can be relieved by the transcription activator LeuO. Mol. Microbiol. *77*, 1380–1393.
White, J.R., Escobar-Paramo, P., Mongodin, E.F., Nelson, K.E., and DiRuggiero, J. (2008). Extensive genome rearrangements and multiple horizontal gene transfers in a population of *Pyrococcus* isolates from Vulcano Island, Italy. Appl. Environ. Microbiol. *74*, 6447–6451.
Whittle, G., Shoemaker, N.B., and Salyers, A.A. (2002). Identification of two genes involved in the modulation of conjugal transfer of the *Bacteroides* conjugative transposon CTnDOT. J. Bacteriol. *184*, 3839–3847.
Wilcks, A., and Jacobsen, B.B. (2010). Lack of detectable DNA uptake by transformation of selected recipients in mono-associated rats. BMC Res. Notes *3*, 49.
Williams, D.W., Kuriyama, T., Silva, S., Malic, S., and Lewis, M.A.O. (2011). *Candida* biofilms and oral candidosis: treatment and prevention. Periodontology *55*, 250–265.
Williams, H.G., Benstead, J., Frischer, M.E., and Paul, J.H. (1997). Alteration in plasmid DNA following natural transformation to populations of marine bacteria. Mol. Mar. Biol. Biotechnol. *6*, 238–247.
Williamson, K.E., Radosevich, M., Smith, D.W., and Wommack, K.E. (2007). Incidence of lysogeny within temperate and extreme soil environments. Environ. Microbiol. *9*, 2563–2574.
Williamson, K.E., Schnitker, J.B., Radosevich, M., Smith, D.W., and Wommack, K.E. (2008). Cultivation-based assessment of lysogeny among soil bacteria. Microb. Ecol. *56*, 437–447.
Wozniak, R.A., and Waldor, M.K. (2010). Integrative and conjugative elements: mosaic mobile genetic elements enabling dynamic lateral gene flow. Nat. Rev. Microbiol. *8*, 552–563.
Wright, M.S., Peltier, G.L., Stepanauskas, R., and McArthur, J.V. (2006). Bacterial tolerances to metals and antibiotics in metal-contaminated and reference streams. FEMS Mircobiol. Ecol *58*, 293–302.
Wright, M.S., Baker-Austin, C., Lindell, A.H., Stepanauskas, R., Stokes, H.W., and McArthur, J.V. (2008). Influence of industrial contamination on mobile genetic elements: class 1 integron abundance and gene cassette structure in aquatic bacterial communities. ISME J. *2*, 417–428.
Wright, G.D. (2010). Antibiotic resistance in the environment: a link to the clinic? Curr. Opin. Microbiol. *13*, 589–594.
Yen, H.C., Hu, N.T., and Marrs, B.L. (1979). Characterization of the gene transfer agent made by an overproducer mutant of *Rhodopseudomonas capsulata*. J. Mol. Biol. *131*, 157–168.
Xiao, R.Y., Wang, Z., Wang, C.X., Yu, G., and Zhu, Y.G. (2006). Genotoxic risk identification of soil contamination at a major industrialized city in northeast China by a combination of *in vitro* and *in vivo* bioassays. Environ. Sci Technol. *40*, 6170–6175.
Young, H.K. (1993). Antimicrobial resistance spread in aquatic environments. J. Antimicrob. Chemother. *31*, 627–635.

Zhang, L.H., and Kerr, A. (1991). A diffusible compound can enhance conjugal transfer of the Ti plasmid in *Agrobacterium tumefaciens*. J. Bacteriol. *173*, 1867–1872.

Zhao, J.S., Deng, Y., Manno, D., and Hawari, J. (2010). *Shewanella* spp. genomic evolution for a cold marine lifestyle and in-situ explosive biodegradation. PLoS One 5, e9109.

Zhu, Y., Li, D., Wang, F., Yin, J., and Jin, H. (2004). Nutritional assessment and fate of DNA of soybean meal from roundup ready or conventional soybeans using rats. Arch. Anim Nutr. *58*, 295–310.

What Maintains Plasmids Among Bacteria?

7

Francisco Dionisio, Teresa Nogueira, Luís M. Carvalho, Helena Mendes-Soares, Sílvia C.M. Mendonça, Iolanda Domingues, Bernardino Moreira and Ana M. Reis

Abstract

The ubiquity of plasmids in nature contrasts with our ability to understand their maintenance. Despite the ability of plasmids to transfer across different bacterial taxonomic groups and to carry useful genes to bacterial cells, it is unclear which factors are responsible for plasmid maintenance among bacterial populations. In this chapter, we present several hypotheses aiming at explaining plasmid existence: efficiency of self-transfer, advantageous genes, transitory derepression of conjugative pili synthesis, compensatory mutations, the existence of amplifier strains, positive epistasis between chromosomal mutations and plasmids, selective sweeps, frequent cross-species transfer, as well as three types of social interactions (exploitation avoidance in the production of public goods, pathogen- or parasite-mediated harmful behaviour, biofilm formation). These hypotheses imply that plasmids and their hosts are adaptable to variable conditions and even that plasmids can be irreplaceable under particular circumstances.

Introduction

Prokaryotes (bacteria and archaea) often harbour plasmids – extrachromosomal molecules of DNA that are present in one or more copies per bacterial cell. Plasmids (together with transposons, bacteriophage elements, and self-splicing molecular parasites) are commonly referred to as mobile genetic elements because they are composed of DNA that can move around within or between genomes (Siefert, 2009). The total of all mobile genetic elements in a cell is referred to as the mobilome (Frost *et al.*, 2005; Siefert, 2009). These genetic elements move between microbial cells, consequently allowing for horizontal transfer of traits in addition to those strictly essential for their replication (Frost *et al.*, 2005).

This chapter focuses on plasmids and the conditions for their maintenance. Most of the examples given will be from enterobacteria, although the arguments presented might apply to other taxonomic groups as well. Enterobacteria are one of the best studied groups of prokaryotes, partly because of their medical and ecological importance. As we will see below, plasmids play a significant role in these contexts.

Plasmids are present in many bacterial strains. For example, in a survey of 202 *Escherichia coli* strains, obtained from 81 American and Australian mammalian species (representing 39 families and 14 orders), Souza and co-workers found one or two types of plasmids per isolate (Souza *et al.*, 1999). In another survey, 140 out of 223 bacterial isolates belonging to seven Enterobacteriaceae species harboured one or more plasmids (Sherley *et al.*, 2003). But not all plasmids are able of transferring to other cells. In a recent study using genomic data, about half (52%) of all proteobacterial plasmids were estimated to be transmissible, but only one-quarter (28%) were classified as conjugative (Smillie *et al.*, 2010).

Plasmid-mediated conjugation is a replicative process during which a plasmid is transferred from a donor to a recipient cell, requiring donor-recipient contact. Sometimes (but rarely) chromosomal DNA is also transferred during this process, as observed by Lederberg and Tatum, when the first plasmid was reported (Lederberg and Tatum, 1946). Contact between bacterial cells is promoted by structures – 'sex pili' – whose proteins are coded in the plasmid DNA. After plasmid transfer, the recipient cell also becomes a plasmid donor. Some plasmids, called 'mobilizable plasmids', are mobilized to other cells solely when a conjugative plasmid is present in the same cell. One of the most impressive facts about conjugative plasmids is that donor and recipient cells can belong to different genera, or even to different kingdoms (Heinemann and Sprague, 1989; Amábile-Cuevas and Chicurel, 1992; Lessl and Lanka, 1994; Courvalin *et al.*, 1995).

When referring to plasmid prevalence and maintenance among bacterial populations, two very different and somewhat contrasting causes are usually given: (i) mutualism – plasmids are maintained because they carry useful genes, hence establishing a mutualistic relationship with their hosts; (ii) parasitism – the transfer rate compensates for its costs, and thus plasmids behave like normal parasites. As we shall see, these two causes are not sufficient to explain plasmid ubiquity.

This chapter has two main parts. We start with the section 'Why plasmids should not exist', where we explain why the two hypotheses mentioned above are not sufficient to explain plasmid maintenance. Then, in the section entitled 'Why plasmids should exist', we present different hypotheses for the maintenance of plasmids. These arguments are primarily of two types: based on individual selection (either of plasmids or of plasmid-bearing bacterial cells) or on social evolution.

We shall discuss different ecological, genetic and evolutionary factors promoting (or not) plasmid maintenance among bacterial populations in nature.

Why plasmids *should not* exist

Innovations usually mean bad news to well adapted living beings. There are two main causes for this maxim. First, the genomic mutation rate that generates beneficial mutations is much lower than the mutation rate that generates deleterious mutations (Perfeito *et al.*, 2007). Second, if some change were adaptive, then it should have been fixed in the population many generations ago. With bacterial cells this notion is often corroborated. For example, newly acquired spontaneous mutations conferring resistance to a given antibiotic are often disadvantageous to its carrier, unless, of course, that antibiotic is present in the environment (Schrag and Perrot, 1996; Lenski, 1998; Trindade *et al.*, 2009). The same seems to apply to plasmids.

Plasmids are costly

Bacterial cells that recently acquired a plasmid often have a lower division rate than the plasmid-free parental strain. This has been observed both with non-conjugative (Bouma and Lenski, 1988; Modi *et al.*, 1991) and conjugative plasmids (Dahlberg and Chao, 2003;

Dionisio *et al.*, 2005). As expected, these novelties decrease the Darwinian fitness of bacteria. But, by how much? How costly are plasmids?

Fig. 7.1 shows values of plasmid costs obtained by different authors, using different methods. Among seven conjugative plasmids, costs varied between 2.8% and 8% (Dionisio *et al.*, 2005; Silva, 2010). The two non-conjugative plasmids conferred costs of 5.5% (Bouma and Lenski, 1988) and 14.7% (Modi *et al.*, 1991).

Can infectious transfer compensate for plasmid costs?

One obvious explanation for the ubiquity of conjugative plasmids is their ability to be transferred to other bacterial cells as parasites. Indeed, many plasmids are able to transfer across different host species and in different environments (Amábile-Cuevas and Chicurel, 1992; Smillie *et al.*, 2010). For plasmid transfer to be an argument for plasmid maintenance, conjugation rate has to compensate both for plasmid cost and segregational loss of the plasmid. Segregational loss is the emergence of plasmid-free bacterial cells upon division of plasmid-bearing cells. Although plasmids have mechanisms to avoid segregation, these are not perfect and plasmid-free cells have been seen to arise in laboratory conditions at relatively high rates (for example, one plasmid-free cell per 10^3 bacterial divisions). Some plasmids code for toxin–antitoxin systems that kill plasmid-free cells arising from the replication of plasmid-bearing cells (Gerdes *et al.*, 2005; Magnuson, 2007). However, these systems do not prevent competition with already existing plasmid-free cells in the population. Therefore, toxin–antitoxin systems cannot explain maintenance of plasmids either.

Using a mass–action mathematical model, Stewart and Levin demonstrated the existence of a broad set of conditions under which infectious transfer can overcome segregational loss and fitness cost (Stewart and Levin, 1977). According to their model (of a chemostat at

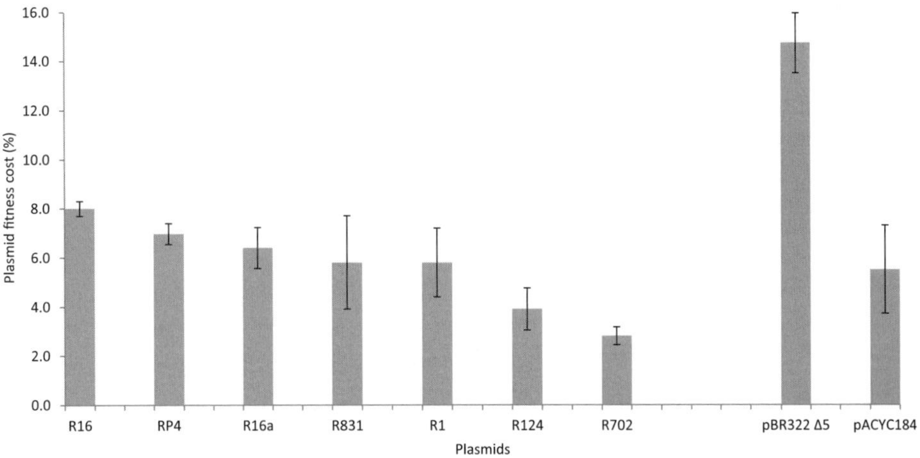

Figure 7.1 The fitness cost of conjugative and non-conjugative plasmids. Data for non-conjugative plasmids pBR322 Δ55 and pACYC184 from Modi and Adams (1991) and Bouma and Lenski (1988) respectively. All the other plasmids are conjugative. Data for R1 from Dahlberg and Chao (2003). Data for the other conjugative plasmids from Silva (2010). Error bars represent standard error.

equilibrium, containing plasmid-free and plasmid-bearing cells, at density N), transfer rate, γ, may explain plasmid maintenance among bacterial populations if $\gamma > (a\rho + s)/N$, where a is the selective advantage of plasmid-free cells, ρ is the rate of turnover of the chemostat and s is the vegetative segregation rate. However, it was later experimentally shown that transfer rates are too low to compensate for segregation loss and costs (Simonsen, 1991; Gordon, 1992). The paradox then remained unresolved.

Two sets of genes are thought to be indispensable for plasmid horizontal spread: the set of mobility genes (MOB genes) – allowing for conjugative DNA processing – and a membrane associated mating pair formation complex (MPF genes) (Smillie *et al.*, 2010). Some plasmids – called mobilizable plasmids – contain all the necessary MOB genes, but lack MPF genes, so they have to 'borrow' these gene products from conjugative plasmids present in the same bacterial cell (Smillie *et al.*, 2010). Therefore, even if conjugation rates were high enough to explain the maintenance of conjugative plasmids among bacterial populations, they could hardly explain the perpetuation of mobilizable and non-conjugative plasmids (Levin and Stewart, 1980). Other plasmids lack both gene complexes: they are neither conjugative nor mobilizable.

Why plasmids *should* exist

In the previous section we expose the major plasmid carriage disadvantage, costs, and how infectious transfer cannot explain their persistence. In the following section, we will see how plasmids and their hosts are adaptive to overcome their limitations and how plasmids can be irreplaceable under particular conditions.

Evolution towards commensalism or mutualism

In theory, although initially costly for bacterial cells, spontaneous mutations occurring either on the plasmid or on the chromosome could compensate for plasmid cost. Compensatory mutations have indeed been observed by several authors, occurring either in the chromosome or in plasmids, or both (coevolution of plasmids and chromosome). Intriguingly, costs disappear in a few hundreds of bacterial generations. This has been observed both with non-conjugative and conjugative plasmids (Bouma and Lenski, 1988; Modi *et al.*, 1991; Dahlberg and Chao, 2003; Dionisio *et al.*, 2005) (see below). Then, compensatory mutations could constitute one of the main solutions for the paradoxical ubiquity of plasmids. We now discuss this in more detail.

Fig. 7.2 shows the outcome of the evolution of different plasmid-bacteria dyads after several hundreds of generations. The evolutionary outcome was similar in all experiments in the sense that plasmid cost disappeared in all of them – but the molecular causes differ. After evolving *E. coli* B harbouring the pACYC184 plasmid for about 500 generations, Bouma and Lenski observed that the cost initially imposed by the plasmid disappeared. However, cells that evolved with the plasmid, when cured of it (i.e. when the plasmid was eliminated from the cell), had a lower fitness than their plasmid-free ancestors: the evolved bacterial cells became addicted to the plasmid (Bouma and Lenski, 1988). The authors then introduced the ancestral plasmid into the evolved cell and the cost was null. This means that the evolved cells were addicted to the plasmid, both in its evolved and ancestral form. Another interesting conclusion from this study is that the main compensatory mutation(s) were present on the chromosome. In another study, also involving a non-conjugative plasmid (pBR322),

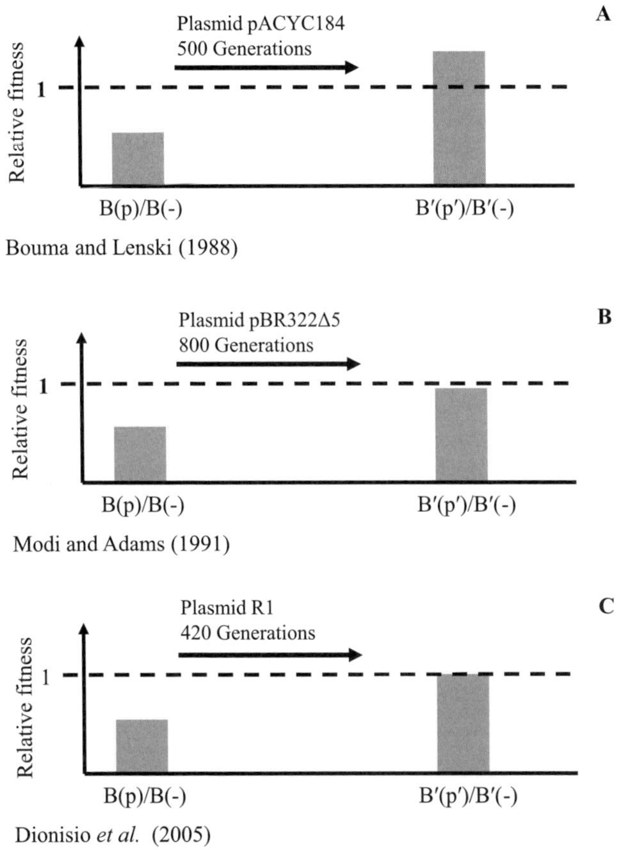

Figure 7.2 The effects of evolved plasmids on the fitness of the evolved *E. coli* hosts, relative to the plasmid-free *E. coli* strain that evolved with the respective plasmid. This is a schematic view of data from (A) Bouma and Lenski (1988), (B) Modi and Adams (1991) and (C) Dionisio *et al.* (2005). In (A) the dyad evolved from parasitism towards mutualism. In (B) and (C) dyads evolved from parasitism to commensalism. *B* denotes the ancestral bacterium genotype, *p* and *p′* denote the ancestral and evolved plasmids and *B′* denotes the genotype that evolved with the plasmid. *B(–)* and *B′(–)* denote plasmid-free clones. Ratios denote relative fitness; for example, *B(p)/B(–)* denotes the fitness of the plasmid-bearing ancestral clone relative to the fitness of the ancestral plasmid-free clone. The clone *B′(–)* was obtained by curing a clone of *B′(p′)*.

Modi and Adams showed that compensatory mutations appeared in both the plasmid and the chromosome, a case of coevolution of the dyad (Modi and Adams, 1991). To account for the fact that conjugative plasmids that evolved in a given host can then move to new hosts, eventually carrying compensatory mutations, Dionisio and co-workers performed evolutionary experiments using the IncFII R1 conjugative plasmid during 420 bacterial generations (Dionisio *et al.*, 2005). Because *E. coli* cells commonly live in structured habitats and in competition with several other strains, this evolution experiment was done in a Petri dish where two strains co-existed whereas Bouma and Lenski (1988) and Modi and Adams (1991) performed their evolutionary experiments in liquid, i.e. unstructured, media and with isolated clones. Dionisio and colleagues reported that the evolved plasmid was not costly to evolved cells (Fig. 7.2), that is, the fitness of the evolved dyad was not significantly different from the fitness of the evolved plasmid-free cell (Dionisio *et al.*, 2005). Moreover, the fitness of the evolved bacteria, harbouring the ancestral plasmid, was also not significantly different from the fitness of the plasmid-free evolved cells. These latter results are important because they mean that, after adaptation to a certain plasmid (with which the chromosome evolved), if the cell looses the evolved plasmid and gains an 'ancestral' one, this may be no longer costly.

Dahlberg and Chao evolved bacterial clones containing conjugative plasmids, either the IncFII R1 plasmid or the IncP RP4 plasmid (Dahlberg and Chao, 2003). After 1100 generations, most dyads evolved towards mutualism (Fig. 7.3), which means the fitness of the evolved dyad is slightly higher than the fitness of the plasmid-free ancestral strain. Similarly, the fitness of the evolved bacterial cell harbouring the ancestral plasmid is also higher than the ancestral plasmid-free cell (Dahlberg and Chao, 2003). Equivalent results were obtained by Dionisio and co-workers (2005) upon evolution of the R1 plasmid in a Petri dish for about 420 generations (Fig. 7.3).

When Dahlberg and Chao (2003) allowed plasmid transfer into the ancestral (plasmid-free) cells, the cost appeared again: the fitness of the ancestral bacteria harbouring the evolved plasmid is lower than the fitness of the ancestral plasmid-free bacterial cells. However, the results of the equivalent experiment by Dionisio's team were different (Fig.

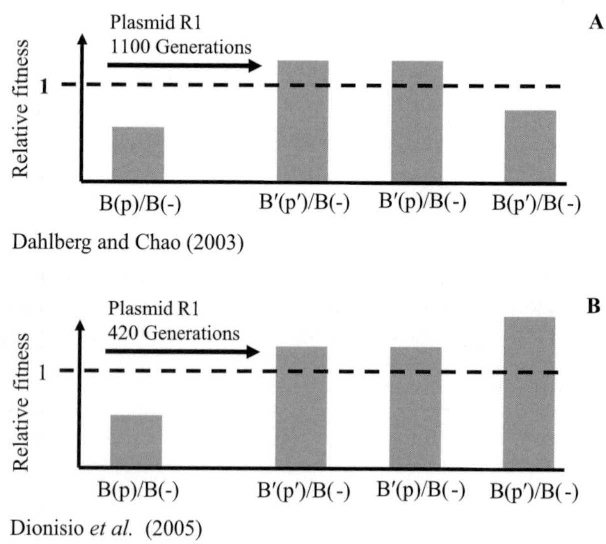

Figure 7.3 The effects of evolved and ancestral plasmids on the fitness of the evolved or the ancestral *E. coli* hosts, relative to the fitness of the ancestral plasmid-free *E. coli* strain. This is a schematic view of data from: (A) Dahlberg and Chao (2003) for the evolution of the R1 plasmid in unstructured (liquid) environment; and (B) Dionisio *et al.* (2005) for the evolution of the R1 plasmid in structured (Petri dish) environment with competition with a second strain. *B*, *B′*, *p*, *p′*, and ratios [e.g. *B′(p′)/B(–)*] as in Fig. 7.2. The clone *B′(p)* results from the introduction of the ancestral plasmid into the cured evolved strain. In both works the evolved R1 plasmid became advantageous to the evolved bacterial cell, relative to the ancestral plasmid-free bacterial cell (second and third bar in A and B). Interestingly, however, in one of the works, when the evolved plasmid moved into ancestral *E. coli* cells, their fitness increased dramatically (Dionisio *et al.*, 2005), a phenomenon not observed in the other work involving the R1 plasmid (Dahlberg and Chao, 2003).

7.3): when the ancestral bacterial cell received the evolved plasmid, its fitness increased considerably (Dionisio *et al.*, 2005). Dionisio and co-workers further allowed the plasmid to be transferred into a different species, *Salmonella enterica* serovar Typhimurium LT2. Although the ancestral plasmid was costly to this species, the evolved plasmid was not. These results are self-explanatory: costly plasmids were no longer so and were now carrying some advantageous mutation to new hosts. This is even more interesting because these experiments were performed with the R1 plasmid, a naturally occurring plasmid that confers resistance to six antibiotics (Dionisio *et al.*, 2005).

When plasmids are not (so) costly to their hosts

In a recent work in our research group (Silva, 2010; Silva *et al.*, 2011) we studied epistatic interactions between five natural conjugative plasmids (all of them carrying resistance genes) and ten chromosomal mutations conferring resistance to the quinolone nalidixic acid (*gyrA* gene, involved in DNA replication), the rifamycin rifampicin (*rpoB* gene, involved in transcription), and the aminoglycoside streptomycin (*rpsL* gene, involved in translation), making a total of 50 different combinations of chromosomal mutations and conjugative plasmids. In the absence of epistatic interactions, the cost for a bacterium to carry both the chromosomal mutation and the plasmid would be the sum of the costs of the mutation and the plasmid. Interestingly, however, 26 out of these 50 combinations (52%) showed positive epistasis, that is the fitness cost of these strains (that carry both resistance determinants) is lower than the independent sum of the cost of each determinant. Therefore, the cost of bearing the two resistance determinants is not as high as one would expect. Moreover, 20 of the 26 combinations presenting positive epistasis constituted instances of sign epistasis: in these cases, the fitness of the cell bearing both the mutation and the plasmid is higher than the fitness of the isogenic cell containing only the mutation or only the plasmid. In other words, the acquisition of an additional resistance determinant (i.e. a plasmid or a mutation) improves the initial fitness cost of the other resistance determinant. The 20 occurrences of sign epistasis represent special cases of compensatory mutations – special because they involve a particular case of mutations, antibiotic resistance mutations – but one can imagine that there are many other possible chromosomal mutations capable of compensating the fitness cost of plasmids.

Finally, the fitness of 3 out of the 50 *E. coli* clones, bearing both a plasmid and an antibiotic resistance mutation, is not different from the fitness of the wild-type *E. coli* strain (without antibiotic resistance mutations nor plasmids). In other words, these three clones may constitute a reservoir of conjugative plasmids (Silva, 2010; Silva *et al.*, 2011).

Plasmids behave as infectious parasites

As explained above, Gordon (1992) has shown that plasmid transfer rates are not high enough to explain their maintenance as parasites only (Simonsen, 1991; Gordon, 1992; Levin, 1993). This conclusion was taken based on a study involving measurement of transfer rates of the R1 plasmid between ten *E. coli* strains. Although it is still unclear why (but see Anderson, 1968; Dionisio, 2005) plasmids, including R1, often repress their own transfer. However, some plasmids transitorily derepress conjugative pili synthesis (Lundquist and Levin, 1986). This happens in cells that just received the plasmids and that transfer them at much higher rates than those that have been carrying them for some time, or those descending from lineages that have carried them for some generations (Lundquist and Levin, 1986).

In 2002, Dionisio and co-workers used the R1 plasmid and a set of nine recently isolated *E. coli* strains as well as four laboratory strains of enterobacteria (*Erwinia chrysanthemi, E. coli, Escherichia fergusoni, Escherichia blattae*) to measure plasmid transfer rates (Dionisio *et al.*, 2002). The authors found that transfer rates span more than six orders of magnitude, as had been observed before with the same plasmid (Gordon, 1992). However, the most surprising finding of this study is that one of the nine *E. coli* strains and the *E. chrysanthemi* strain transferred the plasmid at a very high rate, not only to plasmid-free cells of the same strain but also to other strains, even to strains belonging to other bacterial species.

The discovery that 2 out of 13 enterobacterial strains have extremely high donor abilities has an important consequence. Suppose that among a million cells of a given bacterial strain 'A', a single bacterium contains the R1 plasmid. This cell can, of course, donate the plasmid to other cells. However, as explained above, plasmid transfer rates are usually too low to compensate for plasmid cost (Gordon, 1992; Levin, 1993; Dionisio *et al.*, 2002), meaning that this plasmid-bearing cell would loose in competition with plasmid-free cells. Now suppose that these cells A, among which just a few contain the conjugative plasmid, live among bacteria from another strain or species, 'B'. It is likely that, sooner or later, a B cell receives the plasmid from one of the plasmid-bearing A cells. Finally, if we assume that the transfer rate of that plasmid is very high amongst the B cells, it will largely compensate for the cost of the plasmid among B cells and for its segregation rate. If so, one can infer that the transfer rate from B cells towards A cells is relatively high, in particular when the density of B cells is high. If the transfer rate from B to A cells is high, then a large proportion of A cells may receive the plasmid back from B cells.

The process just described shows that, even if the transfer rates between most bacterial strains are low, an 'amplifier' strain (B strain in the example above) may be enough to transfer the plasmid to cells belonging to other strains.

As mentioned, amongst 13 enterobacterial strains studied (constituting four species), Dionisio and co-workers (2002) found two amplifier strains. Surprisingly, however, they belong to different species: one of the amplifiers is an *E. coli* strain and another one is the *E. chrysanthemi* strain (Dionisio *et al.*, 2002).

One can approach amplification from a different perspective. Consider three bacterial populations, X(p), Y, and Z, living in the same habitat (Fig. 7.4). Strain X(p) is the initial carrier of the plasmid (p) with low transfer ability towards any bacterial cell, Z is the amplifier strain with high donor ability; and Y is the rest of the community (initially, Y and Z cells are plasmid-free). Why should X cells bear the plasmid p? One possibility is that, in another habitat an antibiotic was used, eventually selecting for X(p) cells because the p plasmid confers resistance to that antibiotic. Because of the low donor ability of X(p) cells, the plasmid will propagate to Y recipient cells at a very slow rate. If X(p) and Z cells coexist, X(p) will also donate plasmids at a low rate to Z cells. However, even at this low rate, the high donor ability of Z(p) cells will increase plasmid spread exponentially to the rest of the community. With such amplification effect, a massive plasmid transfer to the rest of the bacterial community is expected. This was observed experimentally with the R1 plasmid (Dionisio *et al.*, 2002), (Fig. 7.5).

According to the 'amplifier' hypothesis (Dionisio *et al.*, 2002) (Fig. 7.4) most transfer rates of a certain plasmid are too low to explain its spread as a parasite only. Only a few strains, the 'amplifier strains', are very potent donors.

The existence of a few amplifier strains among bacterial populations may have an impact

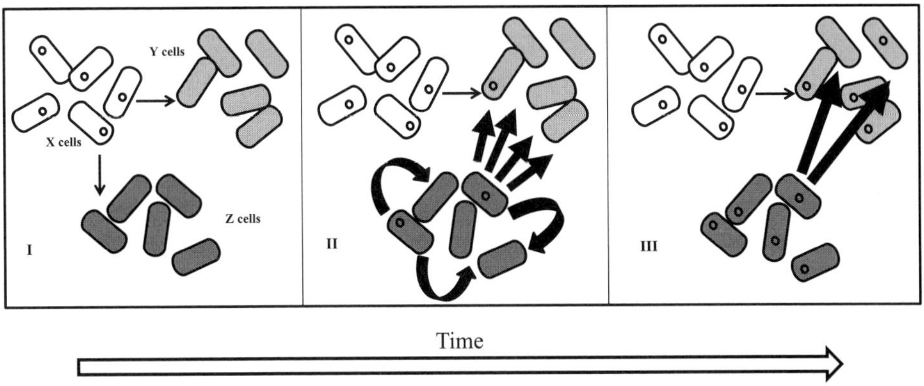

Figure 7.4 The amplification effect. X, Y and Z represent three bacterial strains. Rod-shaped figures represent bacterial cells and small circles inside them represent conjugative plasmids. Thin and thick arrows connecting bacterial cells represent low and high plasmid transfer rates respectively. Initially, all plasmids in the system are carried by X cells, bad donors of the plasmid. Y cells that just received the plasmid are bad donors as well. However, as soon as a Z cell receives the plasmid (e.g. from X cells), transfer among Z cells is intense, as well as from Z towards Y cells. Without Z cells (an amplifier strain because cells are good plasmid donors), most Y cells would remain plasmid-free.

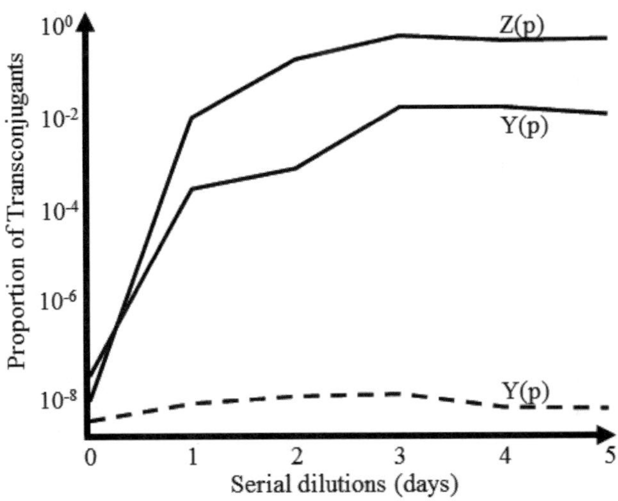

Figure 7.5 Experimental results illustrating the amplification effect of the R1 plasmid. X, Y and Z as in Fig. 7.4. *Y(p)* and *Z(p)* represent plasmid-bearing Y and Z cells respectively. Figure and data adapted from (Dionisio *et al.*, 2002). Plasmid-bearing X cells were grown together with Y cells. After five days, we observed almost no *Y(p)* cells (broken line). When donor X cells (X cells with plasmid p) and Y cells were grown together with amplifier plasmid-free Z cells, many Z and Y cells received the R1 plasmid (full line).

on bacterial population genetics, so complete information about plasmid transfer rates would be extremely useful. Unfortunately, and to our knowledge, there is no systematic study about plasmid transfer rates in nature. Although there are now some biogeographic and metagenomic studies exploring the spatial distribution of bacterial species at different scales on several kinds of microbial communities (Cho and Tiedje, 2000; Vos and Velicer, 2006, 2008; Huber *et al.*, 2007), we are still far from having a clear idea of which type of contact network and, therefore, which type of population structure bacteria really form. There is little doubt, however, that the type of network will influence the conditions for the maintenance of plasmids.

The influence of some types of population structure on population genetics and the role of migration in patterns of neutral variability have been established in different contexts. The first model of population subdivision, the island model, was introduced 80 years ago by Sewall Wright (Wright, 1931) and has been used since then as a reference. In Wright's island model, a species is subdivided into subpopulations, called demes. Migrants are exchanged between demes and they are assumed to come with equal probability from any other deme. To some extent, this simple model allows one to equate the influence of structure in epidemiology and evolutionary change. Since Wright's work, other types of population structure and networks have been studied by mathematicians and population geneticists.

Bacterial communities are often very complex, with perhaps billions of cells per gram of soil, faeces, and other habitats. Despite this complexity, a given plasmid does not have easy access to all bacterial species because some strains are very bad plasmid donors or very bad plasmid recipients (Gordon, 1992; Dionisio *et al.*, 2002). So, each plasmid type only comes into contact with a few of the existing bacterial species. Consider a network where each node (representing a deme) is a bacterial strain. Each node is, then, represented by a myriad of cells of a given strain. If a given plasmid can be transferred by conjugation between two strains, then it means that this plasmid makes a potential connection between the nodes constituted by these two bacterial strains. If this transfer rate is high enough to compensate for plasmid cost and for segregation, then one may consider that the donor strain is a good donor, and that the two nodes are in fact connected. However, when a given strain is a poor donor of a given plasmid to another strain, these two nodes are not connected, and, to get there, the plasmid has to 'travel' through an amplifier strain (a third node).

In a network, the characteristic path length is defined as the average minimal distance between all pairs of nodes in the network. Clustering is defined as the probability that two nodes that are connected to a third node are also connected between them. Unlike regular networks, completely random graph networks present low clustering coefficient together with a small characteristic path length. Studies of the structural properties of networks obtained from distinct natural systems show that regular lattices and completely random graphs do not capture their topological properties (Albert and Barabasi, 2002). The properties of the networks of these natural systems are between an ordered lattice and a completely random graph. Some models have been proposed to better describe these systems. Since 1998, a new kind of structured population has attracted the attention of epidemiologists, mathematicians, physicists and sociologists: the so called Watts and Strogatz small-world networks (Watts and Strogatz, 1998). These networks display a small average path length, like completely random graphs, together with a large clustering coefficient, like regular lattices (e.g. the regular structure formed by the atoms or molecules in a crystal). Recent investigations have found these small-world properties in mammalian cerebral cortical networks (Sakata *et al.*, 2005), *Drosophila melanogaster* protein interaction networks (Middendorf *et al.*, 2005), other protein–protein interaction networks (Rice *et al.*, 2005) and metabolic networks (Jeong *et al.*, 2000), to mention just a few.

It turns out that, owing to these 'amplifier' cells, the bacterial world becomes a 'small world' (Watts and Strogatz, 1998): with their 'help' plasmids can travel much more efficiently between different strains and species. For this reason, most bacterial cells receive plasmids, not from other cells of the same strain or at random from any bacterial cell, but mostly from 'amplifier' strains, which are able to increase the plasmid levels within their strain, in the population, and then deliver them to other strains or even to other species. So, the expected path length of a plasmid being transmitted between two bacterial cells (belonging or not to the same strain) becomes shorter (it only has to go through an amplifier strain), while the probability that two donor cells belong to the same strain is very high because, eventually, most plasmid transfers occur in dyads in which donor cells belong to an amplifier strain. This is in agreement with the definition of a 'small world' network of Watts and Strogatz (Watts and Strogatz, 1998). There is, however, an important difference between the Watts and Strogatz network and the one defined by plasmid transfer between bacterial cells: the networks they analysed are such that the connection from a vertex 'A' to a vertex 'B' is as strong as from vertex B towards vertex A, whereas, in general, the transfer rate of a plasmid from a bacterial cell A to another bacterial cell B is not the same as that from B to A.

To address the structure of plasmid networks among bacteria, our group has been using different plasmids to perform a study similar to that of 2002 (Dionisio *et al.*, 2002), as described above. Our preliminary conclusion of these measurements is that the best donors of our previous study using the IncFII R1 plasmid are also the best donors for plasmids belonging to different incompatibility groups. If confirmed, these data point to a small-world structure for these plasmids, due to the existence of amplifier nodes.

Plasmids have advantageous genes

One of the most profusely disseminated ideas about plasmids is that they encode for genes that augment the fitness of their host, e.g. genes for antibiotic resistance, heavy metal resistance, virulence, fermentation of unusual carbon sources, or UV-light resistance. Plasmids may also block phage entry, as is the case of the F plasmid that blocks the entry of the T7 phage into bacterial cells (Molineux *et al.*, 1989; Schmitt *et al.*, 1995; Cheng *et al.*, 2004; Kemp *et al.*, 2004). Another advantageous function commonly encoded in plasmids is the production of allelopathic substances such as bacteriocins – secreted toxins that can kill other strains of bacteria (Riley and Wertz, 2002). Given that most bacteria live in complex habitats composed by many bacterial strains and species, it is expected that these toxins play an important role in the microbial dynamics at both population level and community level (Riley and Wertz, 2002; Hawlena *et al.*, 2010).

Bergstrom and co-workers, however, argued that advantageous genes do not fully explain the existence and maintenance of plasmids. First, many plasmids have no genes useful to their hosts. Secondly, and this is a crucial point, advantageous genes can, in principle be recruited by the bacterial chromosome (Bergstrom *et al.*, 2000). Various mechanisms facilitate the movement of genes between plasmids and bacterial chromosomes. Indeed, structures such as insertion sequences or transposons are able to move genes among plasmids or between plasmids and chromosomes. Moreover, other structures called integrons are able to 'pick' genes and, by site-specific recombination, integrate them downstream of a promoter (Rowe-Magnus and Mazel, 1999, 2002). Sooner or later, plasmid-bearing cells will have to compete with similar bacterial cells that have lost the plasmid after sequestering the beneficial gene onto the chromosome.

Bergstrom and colleagues argue that plasmids 'find themselves in a bind, trapped between carrying genes that are too useful (in which case those genes will be rapidly incorporated into the chromosome) and carrying genes that are too useless (in which case the plasmid-bearing cells will be outcompeted by plasmid-free cells)' (Bergstrom *et al.*, 2000). In their mathematical analysis, Bergstrom and co-workers (2000) argued that selective sweeps could be helpful to plasmid maintenance. Conjugative plasmids can use their capacity for infectious transfer to infect higher fitness bacterial variants. For this process to explain plasmid maintenance, selective sweeps have to occur frequently enough so that the acquisition of the beneficial genes by the chromosome does not occur before the next selective sweep. Another process that can explain plasmid maintenance is that they can transfer genes across strains or species and, in this way, persist beyond an extinction event (Amábile-Cuevas and Chicurel, 1992; Bergstrom *et al.*, 2000; Dionisio *et al.*, 2002).

Running the risk of overstatement, one may say that in 2001 two papers changed the way we look at plasmids (Ghigo, 2001; Smith, 2001) as they stated that plasmids may play an important role in bacterial social interactions. This will be the subject of the rest of this review.

The involvement of plasmids in social relations

One can summarize the preceding sections in two sentences: (1) survival conditions for plasmids are very narrow, but transitory derepression of conjugative pili synthesis (Lundquist and Levin, 1986), compensatory mutations (Bouma and Lenski, 1988; Modi and Adams, 1991; Dahlberg and Chao, 2003; Dionisio *et al.*, 2005), amplifier strains (Dionisio *et al.*, 2002), sign epistasis or reciprocal sign epistasis between chromosomal mutations and plasmids (Carneiro and Hartl, 2010; Silva, 2010), selective sweeps (Bergstrom *et al.*, 2000) (see also Atwood *et al.*, 1951; Lenski *et al.*, 1991; Travisano *et al.*, 1995; Elena *et al.*, 1996) or frequent cross-species transfer (Bergstrom *et al.*, 2000) may explain their maintenance; (2) if plasmids carry genes useful to their hosts, the chromosome may recruit the genes and discard the rest of the plasmid – hence avoiding paying the cost of plasmid carriage. In 2001, however, Smith proposed a novel hypothesis: the 'raison d'être' of plasmids is a social one and, in that case, the entire plasmid has to be preserved. Before explaining Smith's hypothesis, we must first explain some of the basic concepts of social evolution. For more details, we suggest the following publications: (Axelrod and Hamilton, 1981; West *et al.*, 2006a,b, 2007a–c, 2008).

Consider a population of N haploid individuals (e.g. bacterial cells). Each individual has a 'name', which is simply a number from $i=1$ to $i=N$. Suppose that if they do not interact with each other, then their fitness is one. The proportion of the altruism allele in individual i is p_i. Given that individuals are haploid, $p_i=0$ if individual i does not carry the allele for altruism, or $p_i=1$ if it does. A proportion

$$\overline{p} = \sum_{i=1}^{N} \frac{p_i}{N}$$

of the individuals have the altruism allele. Individuals carrying this allele are willing to pay a fitness cost c to cooperate with another individual and increase its fitness by b.

In evolutionary biology, costs and benefits are always translated into a decrease or an increase of the number of descendants or something nearly equivalent such as probability of offspring survival (Axelrod and Hamilton, 1981; West *et al.*, 2006a,b, 2007a–c, 2008). All bacterial cells that have a chance of replicating have two descendants; however, the probability of survival, the probability of having the opportunity to replicate, or the time a cell takes to divide, may change from bacterium to bacterium. Therefore, costs c and benefits b may represent changes in these traits.

Now let us suppose that individual $i = 1$ carries this altruistic allele and so cooperates with individual $i = 2$. Then, the fitness of the individual $i = 1$ decreases by c and that of individual $i = 2$ is increased by b. In what follows, and without loss of generality, let us consider that $b > c$. All N individuals have their fitness changed by $b - c$ /N. Why? Because individual $i = 1$ leaves less descendants (cost c) and individual $i = 2$ has more descendants (benefit b), hence amounting to a benefit or cost of $b - c$, which is divided by all N individuals. Therefore, the fitness of each individual is:

$$\omega_1 = 1 - c - \frac{b-c}{N}$$

$$\omega_2 = 1 + b - \frac{b-c}{N} \tag{7.1}$$

$$\omega_{\substack{i\neq 1\\ i\neq 2}} = 1 - \frac{b-c}{N}$$

According to our hypotheses, individual $i = 1$ carries the altruistic allele, that is, $p_i = 1$, but nothing has been said about individual $i = 2$; it has only been said that, overall, a proportion $\overline{p}$ of individuals in the population carries the altruistic allele. Population genetics tell us that the proportion of individuals carrying the altruistic allele in the next generation, $\overline{p'}$, is given by

$$\overline{p'} = \sum \omega_i \cdot \frac{p_i}{N}$$

(Hartl and Clark, 1997), where ω_i is the fitness of individual i (see above). With some algebra, it can be shown that the proportion of altruistic alleles in the population increases if:

$$\frac{p_2 - \overline{p}}{1 - \overline{p}} \cdot b - c > 0 \tag{7.2}$$

To clarify the meaning of this mathematical equation, suppose that the bacterial cell $i = 1$ codes for and produces an important substance (with a cost of 5% => $c = 0.05$) that increases the fitness of bacterial cell $i = 2$ by 10% (that is $b = 0.1$). If cell 2 also has the altruist gene, then $p_2 = 1$, which implies that the proportion of the altruist gene will increase. If $p_2 = 0$, the proportion of the altruist gene will decrease with the altruistic act of individual $i = 1$. In the latter situation, cell 1 pays a cost c to help cell 2 to replicate (which does not carry the altruist allele); this is a circumstance unfavourable for the altruist gene as well as for the other genes of the individual. On the other hand, if cell 2 also contained the altruist gene, later it might contribute to cell 1 replication as well.

In general, when individual 1 cooperates with another unspecified individual, there is a certain probability that they are related. In the case of bacterial cells, they either harbour the altruist allele or not, but the probability that any cell in the population harbours the allele can go from zero to 1. Therefore, the relatedness, r, between individual 1 and individual k is given by

$$r = \frac{p_k - \bar{p}}{1 - \bar{p}} \qquad (7.3)$$

Therefore, Equation 7.2 reads: $r \cdot b - c > 0$. This equation is known as Hamilton's rule (Hamilton, 1964b). According to the theory of inclusive fitness, as proposed by Hamilton in 1964, an altruistic gene can only spread in a population if the genetic relatedness multiplied by the actor's benefit overcomes the cost. This is stated in Equations 7.2 and 7.3. In general, population structure and relatedness influence social traits (Taylor, 1992a,b; Wilson *et al.*, 1992; Queller, 1994; West *et al.*, 2002).

Now, we focus on bacterial cells. In principle, if in a small population only one bacterium produces a public good (any useful secreted substance utilizable by neighbours, e.g. siderophores), all its neighbours will benefit from it. This public good is a social cooperative trait coded by the altruistic allele. This situation creates a conflict known as 'tragedy of the commons' (Hardin, 1968; Dionisio and Gordo, 2006). However, if the cooperative gene is carried on a plasmid (or other MGE), it can spread among the population, generating a cooperative population of producers, where all individuals will express the same cooperative trait, and share a high genetic relatedness (West and Gardner, 2010). This can be seen in Equation 7.3, as follows: if bacterium 1 interacts with bacterium k, then they are close to each other. Therefore, a plasmid can self-transfer from bacterial cell 1 to bacterial cell k (by replication, hence leaving a copy in the original cell). If this plasmid carries the gene necessary for the production of the public good, p_k changes from an uncertain number between 0 and 1 to $p_k = 1$ because the bacterial cell received the plasmid. In this case relatedness becomes one.

Plasmids and the social evolution of bacteria

In his work on social evolution of bacterial pathogenesis, Jeff Smith has proposed the cheater hypothesis for maintenance of virulence-factor genes on mobile genetic elements (MGEs), like plasmids (Smith, 2001). According to his hypothesis, in a pathogenic population producing an extracellular virulence factor, non-producer mutant cheaters will increase in frequency because they do not pay the metabolic cost of producing the virulence factor. However, horizontal gene transfer (HGT) can convert them into producers, increasing their genetic relatedness. In this way, virulence factor genes would be maintained on MGEs in order to prevent or overcome cheating (Smith, 2001). More mobile loci generate higher relatedness by gene gain due to HGT, and therefore allow the maintenance of more cooperative traits via kin selection, while gene loss or migration creates unrelated cheaters (Fig. 7.6).

Many of the genes that are carried by plasmids, and MGEs in general, code for proteins that are targeted to be released into the extracellular milieu, and thus are considered public

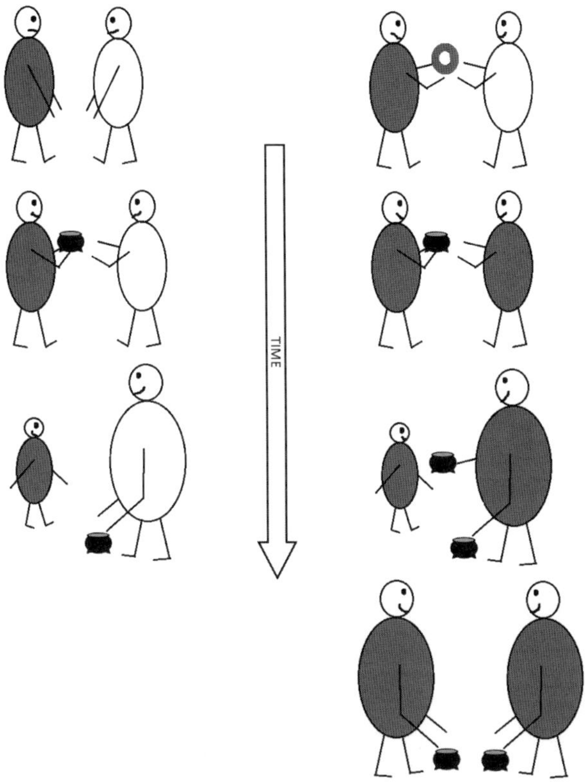

Figure 7.6 The role of plasmids in altruism and kin selection. This figure is based on a figure from Wilson (2000). Two bacterial populations have been reduced to two subjects. Grey subjects contain the altruistic gene (unlike white subjects) that codes for a public good – the pots. The size of subjects represents the fitness of bacterial populations. The doughnut-shaped figure represents conjugative plasmids. Left: grey subjects are not able to pass the altruistic gene towards white subjects. Right: grey subjects pass the cooperative gene towards white subjects, obliging them to cooperate towards the production of a public good. After receiving the plasmid, white subjects become grey, meaning that they become of the same 'type' as the donor grey subject. In the end, both subjects produce the public good and both subjects (both bacterial populations) have the same fitness.

goods (Nogueira *et al.*, 2009; Rankin *et al.*, 2011). Such traits are involved in bacterial sociality such as cooperative behaviours. Social bacteria share their secretome (the set of proteins that are, either released to the extracellular milieu, or exhibited at its surface) and this can be essential for foraging, shelter, microbial warfare and virulence (Nogueira *et al.*, 2009).

To address these points, Nogueira and co-workers have analysed the genomes of 20 *Escherichia* and *Shigella* strains, with different origins within the human microbiome, that were either commensals or of different pathogenic types (Nogueira *et al.*, 2009). Using bioinformatic approaches they inferred the subcellular localization of 58% of the pangenome, the full set of genes of the analysed genomes (Medini *et al.*, 2005). The localization can be divided as: cytoplasm, inner membrane, periplasm, outer membrane and extracellular space. This allowed them to characterize the so-called secretome, as the set of proteins that either belong to the outer membrane or that are extracellular. The analysis of the secretome revealed a close co-occurrence with MGEs, or within mobile regions of the chromosomal genome. The genes of the secretome not belonging to the core genome – the set of genes present in all genomes (that are not expected to be coded elsewhere than on the chromosome) – are found on plasmids rather than on the chromosome over three and a half times more often than would be expected by Pearson's chi-squared test with Yates' continuity correction (with *P*-value < 0.0001). The same tendency does not apply for the other intracellular localizations (Fig. 7.7). This has led to the conclusion that genes coding for cooperative traits are preferentially coded in MGEs, and thus gene mobility drives bacterial cooperation.

On the other hand, cooperation can be maintained by enforcement, as is the case of the co-localization with toxin/antitoxin and restriction-modification systems (Nogueira *et al.*, 2009). They have also classified the secretome according to ancestry (Touchon *et al.*, 2009) and again, they have confirmed that the genes coding for the secretome are frequently lost and gained, supporting the prediction that gene mobility drives bacterial cooperation. We would expect that the proteins of the secretome are costly, rendering populations of cooperators vulnerable to exploitation by non-producing cheaters arising by gene loss or migration. But the computation of the biochemical metabolic cost (according to Akashi and Gojobori, 2002) of all proteins of the pangenome led to the conclusion that the proteins of the secretome are biosynthetically cheaper than the other proteins, even than highly expressed proteins, suggesting selection to lower the cost of cooperative traits (Nogueira *et al.*, 2009).

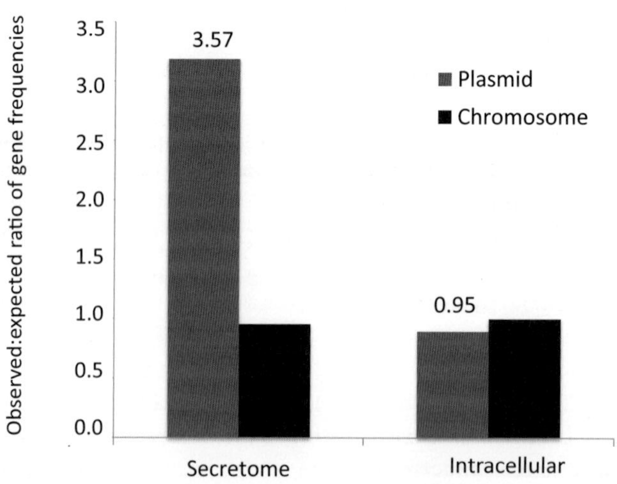

Figure 7.7 Localization of genes coding for the secretome either on plasmids or on chromosomal genomes. The expected ratios of the frequencies of the genes not belonging to the core genome and for whose proteins there is a subcellular localization prediction in a set of 20 *Escherichia* and *Shigella* genomes are observed (Nogueira *et al.*, 2009). The genes were classified in two categories: secretome (those coding for extracellular and outer membrane proteins) and intracellular (those coding for cytoplasmic, inner membrane and periplasmic proteins). The numbers above the bars represent the ratios between gene position frequencies, either on plasmids or on the chromosomal genome. The differences are highly significant (χ^2 test, *P*-value$<$2.2e-16).

The cheater hypothesis for maintenance of virulence-factor genes on MGE, proposed by Smith (2001) assumes that all plasmids carry the cooperative trait. Very recently, Mc Ginty and co-workers extended this model by taking into account the plasmid point of view in the social dilemma of public good exploitation; plasmids not coding for the public good could also be present in the population and out-compete the producer plasmids (Mc Ginty *et al.*, 2010). Their results suggest that cooperation, based on plasmid genes, is not evolutionary stable in a well-mixed, unstructured population, contrary to Smith's suggestion. In the long-term, the non-cooperative plasmids will outcompete the cooperative plasmids. In a spatially structured population, however, the results indicate that cooperation can persist and be evolutionary stable, since structuring increases relatedness (Nogueira *et al.*, 2009; Mc Ginty *et al.*, 2010). Although the cooperation persists, both when the cooperative trait is coded on a plasmid, or chromosomally, Mc Ginty's model proposes that plasmid localization will be selected over the chromosomal one (Mc Ginty *et al.*, 2010). This is in accordance with the findings of Nogueira and colleagues (2009), strengthening the idea that there is an advantage for cooperative traits to be coded on MGEs.

In 2007, West and colleagues listed potential bacterial public goods, which included for example siderophores, quorum-sensing molecules, virulence factors and secreted proteins involved in antibiotic resistance (West *et al.*, 2007a). Antibiotic resistance may be thus considered a bacterial social trait, whose genes are frequently coded in MGEs such as conjugative plasmids (Smillie *et al.*, 2010). This may support Smith's (2001) social hypothesis for plasmid persistence in bacterial populations. Social bacteria can release to the milieu molecules that either detoxify the environment by breaking down antibiotic molecules, or assist neighbouring cells in fighting off the antibiotic. Examples are some enzymes coded by antibiotic resistance genes that can be actively secreted or leaked out the cell, such as β-lactamases that break down β-lactam antibiotics (Neu, 1992). These producer cooperative bacteria will be vulnerable to invasion, by competing against neighbouring non-producer cheaters that take the benefit, without paying the cost of the production for the antibiotic resistance trait. Again, as in the case of Smith's virulence factors previously described, by allowing plasmids to carry these antibiotic resistance genes, producer bacteria can use HGT (plasmid transfer) to increase relatedness and enforce cheaters to cooperate in public good production, thus avoiding exploitation under antibiotic selective pressure. Empirical exploration of the antibiotic resistance social dilemma is emerging. It has been shown, by using engineered *E. coli* strains overproducing β-lactamases, that secretion of plasmid-encoded β-lactamase represents a fitness burden to producing cells and is a public good, which can be exploited by sensitive non-producer cells in the neighbourhood (Dugatkin *et al.*, 2005; Clark *et al.*, 2009; Perlin *et al.*, 2009). Sensitive non-producer cells (cheaters) facing ampicillin only survive when growing in the presence of enough β-lactamase producers. These experimental studies provide evidence of altruistic behaviour in antibiotic resistance, but in this case conjugation was intentionally avoided. Therefore, to experimentally address the social hypothesis of Smith (2001) and Nogueira and co-workers (2009), it will be necessary to check whether HGT reduces the fitness burden imposed by antibiotic-sensitive cheaters. Another important challenge is to perform experiments in more realistic conditions, particularly testing antibiotic resistance coded on natural plasmids.

In experimental studies with diverse natural conjugative plasmids coding for antibiotic resistance, our group has also found (in preparation) that protection of non-producers (plasmid-free cells) against ampicillin is a function of the number and frequency of cells bearing β-lactamase-producing natural plasmids in the neighbourhood (density- and frequency-dependent protection). However, the occurrence and the level of protection seem to be plasmid dependent. Moreover, carriage of these natural plasmids coding for the production of β-lactamase does not always display a (experimentally detectable) fitness cost over the plasmid-free non-producer cells. These results with natural plasmids raise uncertainties about the universality of antibiotic resistance as an example of a social cooperation trait responsible for plasmid persistence. It is worthwhile noting that antibiotic resistance may not entirely meet the condition of a pure social cooperative trait, in the same way that the production of some virulence factors actually is. β-lactamase secretion may not be essential for the individual producer cell survival, and it will be most likely a side effect, whereas for virulence factors a minimal amount of a public good is necessary for a successful host invasion. Empirical confirmation of the hypothesis of HGT as a promoter of social cooperation in antibiotic resistance is then still lacking.

Plasmids and biofilms

Bacteria forming biofilms seem to be more prone to evade, not only attack by the immune system, but also antimicrobial agents, amoebas, and bacteriophages (Costerton *et al.*, 1999, 2005). Ghigo (2001) showed that many sex pili producing plasmids induce the development of biofilms, in particular when plasmid-free cells are around (Ghigo, 2001; Reisner *et al.*, 2006a,b). Now we ask: could plasmidic genes be recruited into the chromosome and be as useful as before? Apparently, the answer is 'no'.

Consider a bacterial cell expressing sex pili. These structures help cells to adhere to surfaces and stabilize the biofilm structure through cell–cell interactions, eventually forming a layer of bacterial cells (Ghigo, 2001; Reisner *et al.*, 2003, 2006a,b; Ong *et al.*, 2009). With the help of sex pili, attached bacterial cells can recruit new (plasmid-free) bacterial cells to the biofilm (Ghigo, 2001) and build the second layer of cells. However, how can this second layer recruit new cells to build the subsequent layers? It is unlikely that the sex pili encoded by the plasmid at the first layer are still useful to build the third layer of bacterial cells. Hence, it is highly desirable that the second layer receives these plasmids: now this second layer can recruit a third layer. The plasmids are then replicated into the third layer and so on. In this case of biofilm formation, plasmids are a very special 'secreted factor' because they are coding for themselves! In other words, it is advantageous for the first layers of cells that the upper layers recruit more cells, so lower level cells just have to pass them the appropriate tools: a copy of the plasmid encoding the sex pili.

Plasmids as biological weapons

Contrary to what happens in most laboratory experiments, each bacterium normally lives in close contact with a myriad of bacterial cells belonging to several strains and species. For example, it has been estimated that humans carry 10-fold more bacterial cells than eukaryotic (human) cells. Most of these bacteria are in our intestines (400–500 species), skin (ca. 200 species) or mouth (ca. 200 species) (Berg, 1996). These are perfect conditions for the selection of bacterial strains encoding allelopathic substances or any other type of weapon. It has been argued that pathogens and parasites can be used by their hosts as biological weapons (Rozsa, 2000; Dionisio, 2007). Indeed, it has been shown that conditions for the selection of pathogen- or parasite-mediated competitive behaviour are broad (Dionisio,

2007). In what follows we argue that conjugative and mobilizable plasmids may be used by bacteria as biological weapons (Fig. 7.8). Before that, however, let us see the evolutionary conditions for such behaviour to be selected.

As discussed above, according to Hamilton's rule, $r \cdot b - c > 0$, a gene coding for altruism can spread in a population if the genetic relatedness multiplied by the actor's benefit overcomes the cost (Hamilton, 1964b). However, this equation is more general than envisaged by Hamilton himself. Indeed, contrary to what was believed in 1964, relatedness can also take negative values. Accordingly, if r is negative, then b has to be negative for Hamilton's rule to be fulfilled. A negative b means that the actor of the behaviour is being spiteful or selfish, not altruist as before. However, how can r be negative? The actor (that harbours the spiteful allele) has to interact with an individual that has a lower probability of bearing the same allele than the population on average. Suppose that individual $i = 1$ has the spite allele and that he will be spiteful towards individual $i = 2$. If somehow individual 1 can *choose* an individual that, for sure, does not carry the spite allele, then p_2 in Equation 7.2 is zero. Then, such spiteful behaviour is selected if:

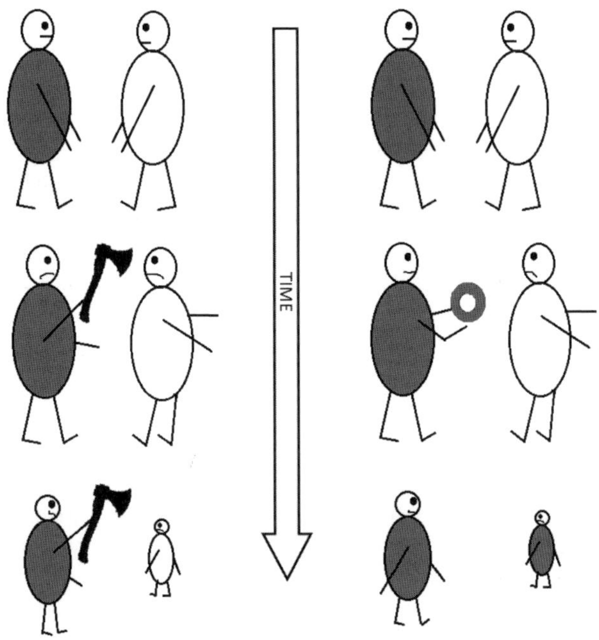

Figure 7.8 Plasmids as spiteful weapons. Figure based on a figure from Wilson (2000). The meaning of subjects and doughnut-shaped figure is as in Fig. 7.6. Two bacterial populations have been reduced to two subjects. Grey subjects contain the spite gene (unlike white subjects). Axes represent a (spiteful) weapon. The size of subjects represents the fitness of bacterial populations. Left: grey subjects are spiteful towards white subjects. Right: grey subjects pass the plasmid towards plasmid-free cells (white subjects), imposing a cost to them. Upon receiving the plasmid, white subjects become grey, as original plasmid donors. However, cells that just received the plasmid, or that descend from a line of cells that received the plasmid a few generations ago, divide more slowly for at least a few hundred generations (Bouma and Lenski, 1988; Modi and Adams, 1991; Dahlberg and Chao, 2003; Dionisio *et al*., 2005). It can be seen that the effect of plasmid transmission is analogous to that of weapon use.

$$\frac{-\bar{p}}{1-\bar{p}}\cdot b - c > 0 \tag{7.4}$$

In the situation just described b is negative because individual 2 is being harmed by the spiteful behaviour of individual 1. Moreover, one can see that

$$r = \frac{-\bar{p}}{1-\bar{p}},$$

which is negative as well.

The theory described by Hamilton (Hamilton, 1964a,b, 1970) and others (Price, 1970, 1972; Grafen, 1985) was mainly built with animal behaviour in mind (but see, for example, Smith, 1976; Axelrod and Hamilton, 1981). However, the application of inclusive fitness theory to plants (Haig and Westoby, 1988, 1989; Queller, 1989) or to the microbial world has proven to be highly successful (Velicer, 2003; West *et al.*, 2007a; Velicer and Vos, 2009). In the case of microbes, one of the reasons is that contact and interactions occur at the molecular level. For example, kin recognition or non-kin recognition may be easily accomplished by molecular signals between bacterial cells. What we now propose is that, possibly, plasmids are used by bacterial cells as biological weapons. This is a direct application of the general hypothesis that hosts may be willing to use their pathogens and parasites to harm other hosts (Rozsa, 2000; Dionisio, 2007).

As we discussed in a previous section, typically plasmids impose a cost to bacterial cells during a short period of time. Later, however, the cell and the plasmid adapt to each other and the cost disappears (Bouma and Lenski, 1988; Modi and Adams, 1991; Dahlberg and Chao, 2003; Dionisio *et al.*, 2005). From this moment on, adapted cells can impose a cost to neighbouring plasmid-free cells by means of passing a copy of their plasmid. This would be more important precisely when cell density is high, which corresponds to the situation in which plasmid transfer is more efficient (Stewart and Levin, 1977; Levin, 1993). Among a population of bacterial cells, plasmid-bearing cells easily transfer their plasmid towards non-kin (in this case negative relatedness), because plasmid-bearing cells have mechanisms to avoid receiving similar or equal plasmids. Let's suppose that a proportion of 90% of the cells bear the plasmid, that is, $\bar{p} = 0.9$; therefore, the relatedness between one of the donors and a recipient cell (10%) is

$$r = \frac{-\bar{p}}{1-\bar{p}} = -9$$

a negative number. If the harm imposed to recipient cells is, for example, 5% = 0.05 of their fitness, then Hamilton's rule tells us that the cost c has to be lower than $-9 \times (-0.05) = 0.45 = 45\%$ for this behaviour (plasmid transfer) to be selected. Data show that only rarely the presence of a plasmid is more than 10% costly (Fig. 7.1). Such cost comprises DNA replication, plasmid transfer, and metabolic burden of plasmid genes, which implies that the cost of donating a plasmid is less than 10%. A cost of 45% for donating a plasmid is clearly above the real cost.

Cryptic plasmids are those that, apparently, do not carry any genes other than those necessary for their own maintenance (and horizontal transmission in the case of conjugative plasmids). By bearing non-conjugative cryptic plasmids, cells avoid being colonized by other plasmids of the same incompatibility group, hence evading spiteful behaviour of neighbouring cells.

Conclusion

Why is it difficult to explain plasmid existence? Clearly, the difficulty does not come from the lack of hypotheses, as can be seen in this review. What is difficult for us is to understand which ones are the most important or most relevant for plasmid maintenance. Probably, different explanations apply to different plasmids. For example, most hypotheses do not apply to cryptic non-conjugative plasmids: these plasmids are maintained, either because they are no longer costly and/or because they are used as shields against colonization by similar (but costly) plasmids. Some plasmids, however, clearly contain genes that are advantageous to the cell. Future research will tell us which other hypotheses are possible and, hopefully, which ones are more relevant for each particular habitat or ecological role.

Acknowledgements

This work was supported by European Science Foundation – EUROCORES – TECT program (Biocontract project) and Fundação para a Ciência e Tecnologia, grants TECH/0001/2007, PTDC/BIA-BDE/66180/2006, post-doc fellowships SFRH/BPD/41863/2007 (TN) and SFRH/BPD/33633/2009 (LMC), and a PhD fellowship SFRH/BD/36813/2007 (HM-S).

References

Akashi, H., and Gojobori, T. (2002). Metabolic efficiency and amino acid composition in the proteomes of *Escherichia coli* and *Bacillus subtilis*. Proc. Natl. Acad. Sci. U.S.A. *99*, 3695–3700.

Albert, R., and Barabasi, A.L. (2002). Statistical mechanics of complex networks. Rev. Mod. Phys. *74*, 47–97.

Amábile-Cuevas, C.F., and Chicurel, M.E. (1992). Bacterial plasmids and gene flux. Cell *70*, 189–199.

Anderson, E.S. (1968). The ecology of transferable drug resistance in the Enterobacteria. Ann. Rev. Microbiol. *22*, 131–180.

Atwood, K.C., Schneider, L.K., and Ryan, F.J. (1951). Periodic selection in *Escherichia coli*. Proc. Natl. Acad. Sci. U.S.A. *37*, 146–155.

Axelrod, R., and Hamilton, W.D. (1981). The evolution of cooperation. Science *211*, 1390–1396.

Berg, R.D. (1996). The indigenous gastrointestinal microflora. Trends Microbiol. *4*, 430–435.

Bergstrom, C.T., Lipsitch, M., and Levin, B.R. (2000). Natural selection, infectious transfer and the existence conditions for bacterial plasmids. Genetics *155*, 1505–1519.

Bouma, J.E., and Lenski, R.E. (1988). Evolution of a bacteria/plasmid association. Nature *335*, 351–352.

Carneiro, M., and Hartl, D.L. (2010). Adaptive landscapes and protein evolution. Proc. Natl. Acad. Sci. U.S.A. *107(Suppl. 1)*, 1747–1751.

Cheng, X.G., Wang, W.F., and Molineux, I.J. (2004). F exclusion of bacteriophage T7 occurs at the cell membrane. Virology *326*, 340–352.

Cho, J.C., and Tiedje, J.M. (2000). Biogeography and degree of endemicity of fluorescent *Pseudomonas* strains in soil. Appl. Environ. Microb. *66*, 5448–5456.

Clark, D.R., Alton, T.M., Bajorek, A., Holden, P., Dugatkin, L.A., Atlas, R.M., and Perlin, M.H. (2009). Evolution of altruists and cheaters in near-isogenic populations of *Escherichia coli*. Front. Biosci. *14*, 4815–4824.

Costerton, J.W., Stewart, P.S., and Greenberg, E.P. (1999). Bacterial biofilms: a common cause of persistent infections. Science *284*, 1318–1322.

Costerton, J.W., Montanaro, L., and Arciola, C.R. (2005). Biofilm in implant infections: its production and regulation. Int. J. Artif. Organs *28*, 1062–1068.

Courvalin, P., Goussard, S., and GrillotCourvalin, C. (1995). Gene transfer from bacteria to mammalian cells. Cr. Acad. Sci. Iii-Vie *318*, 1207–1212.

Dahlberg, C., and Chao, L. (2003). Amelioration of the cost of conjugative plasmid carriage in *Eschericha coli* K12. Genetics *165*, 1641–1649.

Dionisio, F. (2005). Plasmids survive despite their cost and male-specific phages due to heterogeneity of bacterial populations. Evol. Ecol. Res. *7*, 1089–1107.

Dionisio, F. (2007). Selfish and spiteful behaviour through parasites and pathogens. Evol. Ecol. Res. *9*, 1199–1210.

Dionisio, F., and Gordo, I. (2006). The tragedy of the commons, the public goods dilemma, and the meaning of rivalry and excludability in evolutionary biology. Evol. Ecol. Res. *8*, 321–332.

Dionisio, F., Matic, I., Radman, M., Rodrigues, O.R., and Taddei, F. (2002). Plasmids spread very fast in heterogeneous bacterial communities. Genetics *162*, 1525–1532.

Dionisio, F., Conceicao, I.C., Marques, A.C., Fernandes, L., and Gordo, I. (2005). The evolution of a conjugative plasmid and its ability to increase bacterial fitness. Biol. Lett. *1*, 250–252.

Dugatkin, L.A., Perlin, M., Lucas, J.S., and Atlas, R. (2005). Group-beneficial traits, frequency-dependent selection and genotypic diversity: an antibiotic resistance paradigm. Phil. Roy. Soc. Lond. B Biol. *272*, 79–83.

Elena, S.F., Cooper, V.S., and Lenski, R.E. (1996). Punctuated evolution caused by selection of rare beneficial mutations. Science *272*, 1802–1804.

Frost, L.S., Leplae, R., Summers, A.O., and Toussaint, A. (2005). Mobile genetic elements: the agents of open source evolution. Nat. Rev. Microbiol. *3*, 722–732.

Gerdes, K., Christensen, S.K., and Lobner-Olesen, A. (2005). Prokaryotic toxin–antitoxin stress response loci. Nat. Rev. Microbiol. *3*, 371–382.

Ghigo, J.M. (2001). Natural conjugative plasmids induce bacterial biofilm development. Nature *412*, 442–445.

Gordon, D.M. (1992). Rate of plasmid transfer among *Escherichia coli* strains isolated from natural-populations. J. Gen. Microbiol. *138*, 17–21.

Grafen, A. (1985). A geometric view of relatedness. Oxford Surv. Evol. Biol. *2*, 28–90.

Haig, D., and Westoby, M. (1988). Inclusive fitness, seed resources, and maternal care. In Plant Reproductive Ecology: Patterns and Strategies, Lovett Doust, J., and Lovett Doust, L., eds (Oxford University Press New York), pp. 60–79.

Haig, D., and Westoby, M. (1989). Parent-specific gene-expression and the triploid endosperm. Am. Nat. *134*, 147–155.

Hamilton, W.D. (1964a). The genetical evolution of social behaviour II. J. Theor. Biol. *7*, 17–52.

Hamilton, W.D. (1964b). The genetical evolution of social behaviour I. J. Theor. Biol. *7*, 1–16.

Hamilton, W.D. (1970). Selfish and spiteful behaviour in an evolutionary model. Nature *228*, 1218–1220.

Hardin, G. (1968). The tragedy of the commons. Science *162*, 1243–1248.

Hartl, D.L., and Clark, A.G. (1997). Principles of population genetics, 3rd edn (Sinauer Assiciates, Inc. Publishers, Sunderland, MA).

Hawlena, H., Bashey, F., Mendes-Soares, H., and Lively, C.M. (2010). Spiteful interactions in a natural population of the bacterium *Xenorhabdus bovienii*. Am. Nat. *175*, 374–381.

Heinemann, J.A., and Sprague, G.F. (1989). Bacterial conjugative plasmids mobilize DNA transfer between bacteria and yeast. Nature *340*, 205–209.

Huber, J.A., Mark Welch, D., Morrison, H.G., Huse, S.M., Neal, P.R., Butterfield, D.A., and Sogin, M.L. (2007). Microbial population structures in the deep marine biosphere. Science *318*, 97–100.

Jeong, H., Tombor, B., Albert, R., Oltvai, Z.N., and Barabasi, A.L. (2000). The large-scale organization of metabolic networks. Nature *407*, 651–654.

Kemp, P., Gupta, M., and Molineux, I.J. (2004). Bacteriophage T7 DNA ejection into cells is initiated by an enzyme-like mechanism. Mol. Microbiol. *53*, 1251–1265.

Lederberg, J., and Tatum, E.L. (1946). Gene recombination in *Escherichia coli*. Nature *158*, 558–558.

Lenski, R.E. (1998). Bacterial evolution and the cost of antibiotic resistance. Int. Microbiol. *1*, 265–270.

Lenski, R.E., Rose, M.R., Simpson, S.C., and Tadler, S.C. (1991). Long-term experimental evolution in *Escherichia coli.*1. Adaptation and divergence during 2,000 Generations. Am. Nat. *138*, 1315–1341.

Lessl, M., and Lanka, E. (1994). Common mechanisms in bacterial conjugation and Ti-Mediated T-DNA transfer to plant-cells. Cell *77*, 321–324.

Levin, B.R. (1993). The accessory genetic elements of bacteria: existence conditions and (co)evolution. Curr. Opin. Genet. Dev. 3, 849–854.

Levin, B.R., and Stewart, F.M. (1980). The population biology of bacterial plasmids – a-priori conditions for the existence of mobilizable nonconjugative factors. Genetics *94*, 425–443.
Lundquist, P.D., and Levin, B.R. (1986). Transitory derepression and the maintenance of conjugative plasmids. Genetics *113*, 483–497.
Magnuson, R.D. (2007). Hypothetical functions of toxin–antitoxin systems. J. Bacteriol. *189*, 6089–6092.
Mc Ginty, S.E., Rankin, D.J., and Brown, S.P. (2010). Horizontal gene transfer and the evolution of bacterial cooperation. Evolution *65*, 21–32.
Medini, D., Donati, C., Tettelin, H., Masignani, V., and Rappuoli, R. (2005). The microbial pan-genome. Curr. Opin. Genet. Dev. *15*, 589–594.
Middendorf, M., Ziv, E., and Wiggins, C.H. (2005). Inferring network mechanisms: The *Drosophila melanogaster* protein interaction network. Proc. Natl. Acad. Sci. U.S.A. *102*, 3192–3197.
Modi, R.I., and Adams, J. (1991). Coevolution in bacterial-plasmid populations. Evolution *45*, 656–667.
Modi, R.I., Wilke, C.M., Rosenzweig, R.F., and Adams, J. (1991). Plasmid macro-evolution: selection of deletions during adaptation in a nutrient-limited environment. Genetica *84*, 195–202.
Molineux, I.J., Schmitt, C.K., and Condreay, J.P. (1989). Mutants of bacteriophage-T7 that escape F-restriction. J. Mol. Biol. *207*, 563–574.
Neu, H.C. (1992). The crisis in antibiotic-resistance. Science *257*, 1064–1073.
Nogueira, T., Rankin, D.J., Touchon, M., Taddei, F., Brown, S.P., and Rocha, E.P.C. (2009). Horizontal gene transfer of the secretome drives the evolution of bacterial cooperation and virulence. Curr. Biol. *19*, 1683–1691.
Ong, C.L.Y., Beatson, S.A., McEwan, A.G., and Schembri, M.A. (2009). Conjugative plasmid transfer and adhesion dynamics in an *Escherichia coli* biofilm. Appl. Environ. Microb. *75*, 6783–6791.
Perfeito, L., Fernandes, L., Mota, C., and Gordo, I. (2007). Adaptive mutations in bacteria: high rate and small effects. Science *317*, 813–815.
Perlin, M.H., Clark, D.R., McKenzie, C., Patel, H., Jackson, N., Kormanik, C., Powell, C., Bajorek, A., Myers, D.A., Dugatkin, L.A., *et al.* (2009). Protection of *Salmonella* by ampicillin-resistant *Escherichia coli* in the presence of otherwise lethal drug concentrations. Proc. Biol. Sci. *276*, 3759–3768.
Price, G.R. (1970). Selection and covariance. Nature *227*, 520–521.
Price, G.R. (1972). Extension of covariance selection mathematics. Ann. Hum. Genet. *35*, 485–490.
Queller, D.C. (1989). Inclusive fitness in a nutshell. Oxford Surv. Evol. Biol. *6*, 73–109.
Queller, D.C. (1994). Genetic relatedness in viscous populations. Evol. Ecol. *8*, 70–73.
Rankin, D.J., Rocha, E.P., and Brown, S.P. (2011). What traits are carried on mobile genetic elements, and why? Heredity *106*, 1–10.
Reisner, A., Haagensen, J.A.J., Schembri, M.A., Zechner, E.L., and Molin, S. (2003). Development and maturation of *Escherichia coli* K-12 biofilms. Mol. Microbiol. *48*, 933–946.
Reisner, A., Holler, B.M., Molin, S., and Zechner, E.L. (2006a). Synergistic effects in mixed *Escherichia coli* biofilms: Conjugative plasmid transfer drives biofilm expansion. J. Bacteriol. *188*, 3582–3588.
Reisner, A., Krogfelt, K.A., Klein, B.M., Zechner, E.L., and Molin, S. (2006b). *In vitro* Biofilm formation of commensal and pathogenic *Escherichia coli* strains: impact of environmental and genetic factors. J. Bacteriol. *188*, 3572–3581.
Rice, J.J., Kershenbaum, A., and Stolovitzky, G. (2005). Lasting impressions: Motifs in protein-protein maps may provide footprints of evolutionary events. Proc. Natl. Acad. Sci. U.S.A. *102*, 3173–3174.
Riley, M.A., and Wertz, J.E. (2002). Bacteriocins: evolution, ecology, and application. Annu. Rev. Microbiol. *56*, 117–137.
Rowe-Magnus, D.A., and Mazel, D. (1999). Resistance gene capture. Curr. Opin. Microbiol. *2*, 483–488.
Rowe-Magnus, D.A., and Mazel, D. (2002). The role of integrons in antibiotic resistance gene capture. Int. J. Med. Microbiol. *292*, 115–125.
Rozsa, L. (2000). Spite, xenophobia, and collaboration between hosts and parasites. Oikos *91*, 396–400.
Sakata, S., Komatsu, Y., and Yamamori, T. (2005). Local design principles of mammalian cortical networks. Neurosci. Res. *51*, 309–315.
Schmitt, C.K., Kemp, P., and Molineux, I.J. (1995). Streptomycin- and rifampin-resistant mutants of *Escherichia coli* perturb F exclusion of bacteriophage-T7 by affecting synthesis of the F-plasmid protein Pifa. J. Bacteriol. *177*, 1589–1594.
Schrag, S.J., and Perrot, V. (1996). Reducing antibiotic resistance. Nature *381*, 120–121.
Sherley, M., Gordon, D.M., and Collignon, P.J. (2003). Species differences in plasmid carriage in the *Enterobacteriaceae*. Plasmid *49*, 79–85.
Siefert, J.L. (2009). Defining the mobilome. Methods Mol. Biol. *532*, 13–27.

Silva, R.F. (2010). Study of the biological cost of antibiotic resistance. University of Lisbon, Lisbon, Portugal, pp. 50.

Silva, R.F., Mendonça, S.C.M., Carvalho, L.M., Reis, A.M., Gordo, I., Trindade, S., and Dionisio, F. (2011). Pervasive sign epistasis between conjugative plasmids and drug-resistance chromosomal mutations. Plos Genet. *7*, e1000344.

Simonsen, L. (1991). The existence conditions for bacterial plasmids: theory and reality. Microb. Ecol. *22*, 187–205.

Smillie, C., Garcillan-Barcia, M.P., Francia, M.V., Rocha, E.P.C., and de la Cruz, F. (2010). Mobility of plasmids. Microbiol. Mol. Biol. R. *74*, 434–452.

Smith, J. (2001). The social evolution of bacterial pathogenesis. Proc. Biol. Sci. *268*, 61–69.

Smith, J.M. (1976). Evolution and theory of games. Am. Sci. *64*, 41–45.

Souza, V., Rocha, M., Valera, A., and Eguiarte, L.E. (1999). Genetic structure of natural populations of *Escherichia coli* in wild hosts on different continents. Appl. Environ. Microb. *65*, 3373–3385.

Stewart, F.M., and Levin, B.R. (1977). The population biology of bacterial plasmids: a *priori* conditions for the existence of conjugationally transmitted factors. Genetics *87*, 209–228.

Taylor, P.D. (1992a). Altruism in viscous populations – an inclusive fitness model. Evol. Ecol. *6*, 352–356.

Taylor, P.D. (1992b). Inclusive fitness in a homogeneous environment. Proc. Roy. Soc. Lond. B Biol. *249*, 299–302.

Touchon, M., Hoede, C., Tenaillon, O., Barbe, V., Baeriswyl, S., Bidet, P., Bingen, E., Bonacorsi, S., Bouchier, C., Bouvet, O., *et al.* (2009). Organised genome dynamics in the *Escherichia coli* species results in highly diverse adaptive paths. Plos Genet. *5*, e1000344.

Travisano, M., Mongold, J.A., Bennett, A.F., and Lenski, R.E. (1995). Experimental tests of the roles of adaptation, chance, and History in evolution. Science *267*, 87–90.

Trindade, S., Sousa, A., Xavier, K.B., Dionisio, F., Ferreira, M.G., and Gordo, I. (2009). Positive epistasis drives the acquisition of multidrug resistance. Plos Genet. *5*, e1000578.

Velicer, G.J. (2003). Social strife in the microbial world. Trends Microbiol. *11*, 330–337.

Velicer, G.J., and Vos, M. (2009). Sociobiology of the Myxobacteria. Annu. Rev. Microbiol. *63*, 599–623.

Vos, M., and Velicer, G.J. (2006). Genetic population structure of the soil bacterium *Myxococcus xanthus* at the centimeter scale. Appl. Environ. Microb. *72*, 3615–3625.

Vos, M., and Velicer, G.J. (2008). Isolation by distance in the spore-forming soil bacterium *Myxococcus xanthus*. Curr. Biol. *18*, 386–391.

Watts, D.J., and Strogatz, S.H. (1998). Collective dynamics of 'small-world' networks. Nature *393*, 440–442.

West, S.A., and Gardner, A. (2010). Altruism, spite, and greenbeards. Science *327*, 1341–1344.

West, S.A., Pen, I., and Griffin, A.S. (2002). Conflict and cooperation – cooperation and competition between relatives. Science *296*, 72–75.

West, S.A., Gardner, A., and Griffin, A.S. (2006a). Altruism. Curr. Biol. *16*, R482–R483.

West, S.A., Griffin, A.S., Gardner, A., and Diggle, S.P. (2006b). Social evolution theory for microorganisms. Nat. Rev. Microbiol. *4*, 597–607.

West, S.A., Diggle, S.P., Buckling, A., Gardner, A., and Griffins, A.S. (2007a). The social lives of microbes. Annu. Rev. Ecol. Evol. S *38*, 53–77.

West, S.A., Griffin, A.S., and Gardner, A. (2007b). Evolutionary explanations for cooperation. Curr. Biol. *17*, R661–R672.

West, S.A., Griffin, A.S., and Gardner, A. (2007c). Social semantics: altruism, cooperation, mutualism, strong reciprocity and group selection. J. Evol. Biol. *20*, 415–432.

West, S.A., Griffin, A.S., and Gardner, A. (2008). Social semantics: how useful has group selection been? J. Evol. Biol. *21*, 374–385.

Wilson, D.S., Pollock, G.B., and Dugatkin, L.A. (1992). Can altruism evolve in purely viscous populations. Evol. Ecol. *6*, 331–341.

Wilson, E.O. (2000). Sociobiology. The new synthesis, 25th anniversary edition (The Belknap Press, Harvard University Press, Cambridge, MA).

Wright, S. (1931). Evolution in mendelian populations. Genetics *16*, 0097–0159.

Identification of Mobile Genetic Elements in Metagenomes

8

Peter Mullany and Adam P. Roberts

Abstract

Mobile genetic elements are discrete segments of nucleic acid that can translocate from one part of the genome to another, and in the case of conjugative elements between genomes in different cells. The vast majority of our knowledge of mobile genetic elements is derived from experimentation on cultivable bacteria and is therefore incomplete. In this chapter we will discuss the methods available for identifying and isolating these elements from metagenomic samples.

Introduction

Mobile genetic elements (MGEs) are discrete segments of nucleic acid that can translocate from one part of a genome to another, in the case of the mobilizable and conjugative MGEs the donor and recipient genomes can be in different cells. The simplest of the MGEs are insertion sequences (ISs) which only encode the information required for transposition; these elements are not capable of independent conjugation to a new host. ISs can sometimes flank cargo DNA which does not encode any proteins involved in transposition; these more complex elements are called transposons. Transposons are a highly heterogeneous group of mobile genetic elements that transpose to new sites using a variety of mechanisms, a detailed discussion of these is beyond the scope of this chapter but the interested reader is referred to Haniford (2006), Siguier *et al.* (2006), Wagner (2006) and Roberts *et al.* (2008).

Some elements which can transpose are also capable of conjugative transfer. These are called conjugative transposons or integrative conjugative elements (ICEs) (Roberts and Mullany, 2009; Wozniak *et al.*, 2010). These are relatively complex elements which contain regions devoted to conjugation, regulation, integration and excision from genomes and an accessory region which encodes functions not related to mobility.

Plasmids are mobile genetic elements that can replicate independently of the bacterial chromosome. Like transposons these elements are also highly heterogeneous with some being capable of conjugation and others of being mobilized. There have been numerous extensive reviews of the properties of plasmids (e.g. Smillie *et al.*, 2010).

Bacterial viruses or bacteriophage are also considered as MGEs as some of them are capable of transposition as well as lysing their bacterial host and infecting a new host cell (Harshey and Jayaram, 2007).

The mobile introns are also important mobile elements in bacteria as these elements, as well as being able to splice, are capable of transposition to new genomic sites (Toro *et al.*,

2007). These elements are frequently contained within conjugative elements and are widely distributed. They may be involved in modulating the expression of their host gene (Chee and Takami, 2011). Furthermore, they have been used in biotechnological applications to generate mutations in a wide range of different bacterial hosts (Yao and Lambowitz, 2007).

It is important to study MGEs for many reasons, as they have been shown to be important vectors in the spread of antibiotic resistance (Roberts and Mullany, 2011) and bacterial virulence factors (Otto, 2010) and they are central players in the evolution of bacteria. In addition MGEs are used extensively in genetic manipulation of both prokaryotes and eukaryotes (Febrer *et al.*, 2011).

As the majority of microorganisms on and in the planet cannot be cultivated in the laboratory the full contribution of MGEs cannot yet be understood. Therefore various metagenomic approaches have been, and can be, used to identify MGEs from both the cultivable and non-cultivable microbes of any ecological niche. The rest of this chapter describes progress in this area.

The concept of the mobilome

The mobilome is defined as the totality of MGEs in a particular genome. An extension of this idea encompasses the meta-mobilome which can be defined as the totality of MGEs in a particular ecological niche, including MGEs in both the cultivable and non-cultivable organisms. In the rest of this chapter we will talk about what MGEs have been found in the meta-mobilome and the techniques which can be employed to identify them.

Metagenomic studies can be used to isolate mobile genetic elements

There are a variety of different types of assays which can be used to isolate and/or detect novel mobile genetic elements from the metagenome of a bacterial community. These experimental approaches can be broadly divided into four categories: (i) sequencing studies, (ii) sequence-based screening, (iii) capture studies and (iv) functional studies. Each type of approach will be considered in turn.

Sequencing

Next-generation sequencing technologies provide researchers with a fast way to gain an enormous amount of sequence data from the metagenome of a particular niche. Quite often enrichment procedures can be adopted before the isolation of the relevant DNA, e.g. bacteria resistant to various antibiotics have been enriched by culturing them in the antibiotics before isolating and sequencing the plasmid metagenome from wastewater treatment plants (Schlüter *et al.*, 2008; Szczepanowski *et al.*, 2008). In these studies the sequence reads were compared to the environmental gene tags (EGTs) and used to assign a probable function, revealing a high diversity of plasmids from this particular environment. Similar approaches have been used to enrich for viral particles (bacteriophage) before DNA isolation and sequencing of metagenomic DNA from many environments.

In addition metagenomic sequencing projects provide the raw data which can often, after the correct *in silico* assembly and analysis, show the presence of examples of all types of

MGEs. For example, recent analysis of the human gut metagenome and the chicken caecum metagenome revealed genes associated with a wide range of mobile elements (Kurakowa *et al.*, 2007; Qu *et al.*, 2008).

Sequence-based screening

PCR screening

Many MGEs have conserved regions of nucleotide sequence to which PCR primers can be designed in order to amplify any variable intervening sequence. An excellent example of this are the integrons. Each variable gene cassette in an integron is flanked by a conserved region which acts as the substrate for recombination which is catalysed by the integron's integrase. Primers designed from the region (called the 59-bp element) will amplify the intervening regions which are the mobile gene cassettes (Stokes *et al.*, 2001) and therefore this approach has been used in different environments to detect novel gene cassettes.

We have recently tried this approach on oral metagenomic DNA isolated from healthy volunteers across Europe and found an array of different genes which are located between these conserved regions and are therefore believed to be integron cassettes (Fig. 8.1 and Table 8.1), suggesting that integron cassettes are common in the oral cavity.

Library production and screening

Integrated MGEs range from the small ISs to large genomic islands (some greater than 100 kb). In this section we will concentrate on isolation of ICE including conjugative transposons and genomic islands.

ICE elements have a modular organization and typically include a conjugation module (this can be a large complex region in conjugative elements but can be very simple in

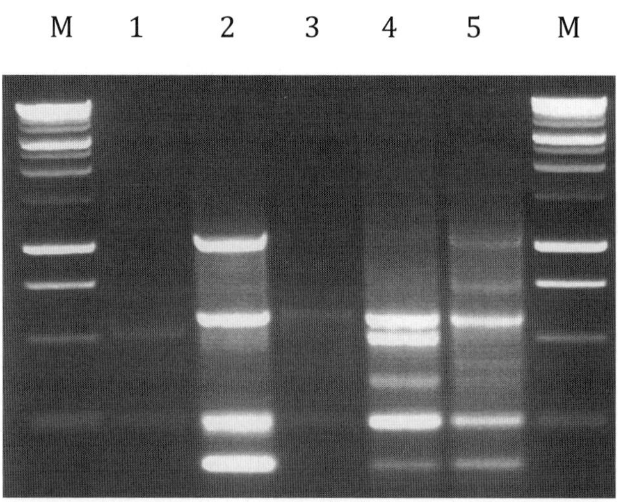

Figure 8.1 A PCR profile of gene cassettes. PCR reactions carried out on human oral metagenomic DNA from different countries. Lane 1, England; Lane 2, Italy; Lane 3, Scotland; Lane 4, Norway; Lane 5, Finland.

Table 8.1 Table showing the closest homologue of some of the amplified gene cassettes and the host of the closest homologue

Closest homologue in Genbank (BLASTX)	Host of closest homologue
Glycerate kinase, putative	*Streptococcus sanguinis* SK36
Uracil phosphoribosyltransferase	*Fusobacterium nucleatum* ATCC 10953
Putative outer membrane protein OmpH	*Bacteroides vulgatus* ATCC 8482
Nicotinate phosphoribosyltransferase	*Neisseria gonorrhoeae* FA 1090
Putative hydrolase, TatD family	*Arthrobacter aurescens* TC1
Multidrug resistance protein, putative	*Vibrio cholerae* AM-19226
3-Oxoacyl-(acyl carrier protein) synthase	*Synechococcus* sp. WH 7805
Translation elongation factors (GTPases)	*Brevibacterium linens* BL2
Elongation factor G	*Arthrobacter* sp.
Beta-N-acetylhexosaminidase	*Bacteroides fragilis* NCTC 9343

mobilisable elements; the latter will just contain an origin of transfer and a *mob* gene). These elements also contain a regulatory region, a recombination region and often accessory regions that are not involved in conjugation or integration and excision from genomes. The accessory region frequently contains antibiotic resistance genes making these elements clinically important. There are a number of recent comprehensive reviews of these elements (Roberts and Mullany, 2009; Frost and Koraimann, 2010; Wozniak and Waldor, 2010).

As most large integrative elements investigated so far have a recombinase gene which mediates integration and excision from the host genome, these can be used as probes with which to screen large insert metagenomic libraries. Most integrative elements have either a tyrosine or a serine recombinase (these refer to the amino acid at the active site of the enzyme; Grindley *et al.*, 2006). Multiple alignments of the tyrosine-based recombinases have allowed the identification of regions that are homologous between most of this family of recombinases (Brouwer *et al.*, 2011) and subsequently these conserved regions have been used as targets for PCR primers in metagenomic DNA. The resulting amplicons are currently being used as probes to screen metagenomic BAC libraries derived from oral metagenomic DNA. The serine-based recombinases are more diverse, therefore more probes to these genes are required (Smith and Thorpe, 2002). As well as isolating ICEs, probing with integrase genes will allow the isolation of integrative phage, some genomic islands and gene capture elements such as integrons.

Tyrosine and serine recombinases are not the whole story when it comes to genomic islands as recently a DDE (DDE refers to the amino acids, in the single-letter code, at the active site of the transposase) type transposase was found associated with a conjugative transposon (Brochet *et al.*, 2009). As these integrative elements are modular in organization, in theory at least, there should be no reason why any recombinase or transposase could not mediate integration and excision.

Hybridization

DNA–DNA hybridization is a very easy method to ascertain if a specific gene or set of genes are present within a metagenomic library. One drawback, which is common to all of the sequence-based approaches discussed here, is that the methodology is based on known sequence and therefore cannot result in knowledge of completely new genes; it can however identify different homologues and further experimentation can determine if these homologues are present on MGEs.

An example of the successful use of hybridization to identify novel putative MGEs is the investigation into magnetotactic bacteria (MTB). The authors initially enriched for MTB, using magnets, creating a metagenomic library from them. Over 5000 clones, with an average insert size of 33 kb were screened by a pool of probes derived from known magnetosome genes revealing one novel magnetosome gene cluster (Jogler *et al.*, 2009). Further metagenomic analysis of uncultivated MTB revealed that the genes responsible for magnetosome formation are encoded on a large genomic magnetosome island and this island is likely to have undergone extensive horizontal gene transfer (Jogler *et al.*, 2011).

Seville *et al.* (2009) analysed human oral and faecal metagenomes by direct hybridization of DNA to probes derived from antibiotic resistance genes which were immobilized on a membrane. The probes were derived from a variety of different antibiotic resistance genes and integrases located on mobilizable and conjugative transposons. This enabled the estimation of the relative abundance of the different genes in the metagenome from these samples. It was found that the conjugative transposon Tn*916* was abundant in oral samples but undetectable in faecal samples and the mobilizable transposon Tn*4451* was equally abundant in faecal samples but absent in oral samples, probably reflecting the likely host of these elements (Seville *et al.*, 2009).

Capture studies

Hybridization with magnetic bead capture

Subtractive hybridization magnetic bead capture is a method developed to allow the capture of relatively long DNA fragments based on their homology to a PCR amplicon which has been bound to a magnetic bead (Meyer *et al.*, 2007).

Using degenerate primers designed from the consensus sequence of the target gene a fragment can be amplified from the metagenomic DNA sample. One of the primers is biotinylated and this facilitates the amplicons immobilization onto streptavidin-coated magnetic beads.

Next, the metagenomic DNA is either mechanically sheared or partially digested with a suitable restriction enzyme. The ends are linked to aptamer oligonucleotides which contain specific binding sites for the final stage primers. The DNA fragments, flanked by the oligonucleotide aptamers are hybridized to the probe bound to the magnetic beads. The DNA that binds to the probe can then be eluted from the beads following stringent washing steps to remove any non-specific background DNA. Lastly the entire length of the hybridizing fragment can be amplified, by primers which bind to the aptamers, cloned and sequenced. One of the benefits of using this method versus PCR based techniques is the high specificity which can be achieved by using high-stringency hybridization conditions. One limitation is that there is no way to ensure capture of a full-length gene as all sequences containing the target may be captured. This methodology has previously been used to isolate large integrons from metagenomic samples (Scotta *et al.*, 2007).

Isolation and properties of plasmids isolated from metagenomes

Like other MGEs plasmids frequently encode properties that allow them to exploit specialized ecological niches, such as antibiotic resistance genes and specialized metabolic pathways such as the utilization of exotic substrates (see below).

The key property of plasmids is that they replicate separately from the bacterial chromosome, and this can be exploited to isolate them from metagenomes. Typically, plasmids are extra-chromosomal elements that contain their own origin of replication and they have been found in almost all bacteria. At their simplest, plasmids just contain an origin of replication and a *rep* gene. Many plasmids also contain an origin of transfer and genes encoding functions that allow them to transfer to new hosts via conjugation (Smillie *et al.*, 2010). Plasmids that contain conjugation genes are called conjugative and those that only contain an origin of transfer (*oriT*) but no conjugation genes are called mobilisable. The latter elements can make use of conjugation functions of conjugative plasmids and/or transposons to transfer to a new host.

As well as circular plasmids, linear plasmids have also been found although they are not as well studied. This probably reflects the relative ease by which circular plasmids can be separated from the bacterial chromosome. However, linear plasmids are now being characterized and have been shown to convey advantageous phenotypes on the host and to be often capable of conjugation (Meinhart *et al.*, 1997; Chaconas and Kobryn, 2010).

Isolation by transposon-aided capture

In this approach no assumptions need to be made about the properties of the plasmids and the plasmid-containing organisms do not need to be cultivable in the laboratory. In this technique metagenomic DNA is prepared from the ecosystem of interest and plasmids are enriched using nucleases that specifically degrade linear DNA. The next step is an *in vitro* transposition reaction in which a Tn5 derivative containing an *E. coli* origin of replication is transposed into the circular DNA molecules (Jones and Marchesi, 2007; Warburton *et al.*, 2011). The transposition reaction is then transformed into *E. coli* selecting for the antibiotic resistance gene on the transposon. We have used this approach to isolate novel plasmids from the oral metagenome (Warburton *et al.*, 2011) and others have used it to isolate novel plasmids from the gut metagenome and from other environmental niches (Jones *et al.*, 2010).

In addition to plasmids, this approach should be able to isolate any circular molecules, with possible size constraints (see below). This is important as many mobile elements, such as conjugative and some non-conjugative transposons and gene cassettes have a circular intermediate form (for some examples see Adams *et al.*, 2002; Roberts and Mullany, 2011). In support of this we isolated a circular molecule that appeared to be generated by a site-specific recombinase rather than being a plasmid (Warburton *et al.*, 2011).

A potential limitation of the transposon-aided capture (TRACA) technology is that only relatively small plasmids have been isolated so far. However, this is a relatively new technology and further optimization such as using different replicons and host–vector systems should overcome this problem. TRACA is not expected to capture linear plasmids as the origin of replication on the Tn5 derivative does not have the ability to replicate their termini; specialized enzymes are required for replication of these elements (reviewed in Ravin, 2003). However, it is conceivable that the TRACA system could be further modified to be able to isolate these types of MGEs.

Functional-based screens

Phenotypic assays

Phenotypic assays represent by far the easiest method of obtaining clones from metagenomic libraries. Many of these screens are based simply on the ability of the host strain to grow in particular conditions. For example, Diaz-Torres *et al.* (2003) screened a human saliva metagenomic library in *E. coli* directly on agar containing 10 μg/ml tetracycline, allowing the isolation of a novel tetracycline resistance gene, *tet*(37). Also a novel conjugative transposon Tn*6079* (de Vries *et al.*, 2011) has been found by screening a faecal metagenomic fosmid library; this demonstrates the power of phenotypic screening.

Another phenotype which has been used to isolate a novel MGE is the degradation of aromatic compounds. Suenaga *et al.* (2009) were able to isolate a plasmid involved in the degradation of extradiol dioxygenase (EDO) from clones selected on EDO. The plasmid also harboured a novel transposon, Tn*6032*.

Concluding remarks

Our understanding of MGEs has to a large part been limited to those that confer clinically important traits such as antibiotic resistance. Other potential MGEs have been identified by whole genome sequencing. However, to gain a comprehensive understanding of MGEs the elements themselves need to be isolated so that we can understand the mechanisms underlying their mobility and understand the properties they confer on their hosts. A range of different but not mutually exclusive metagenomics techniques are available to the experimenter so that a comprehensive understanding of MGE in metagenomes is obtained.

References

Adams, V., Lyras, D., Farrow, K.A., and Rood, J.I. (2002). The clostridial mobilisable transposons. Cell Mol. Life Sci. *59*, 2033–2043.

Brochet, M., Da Cunha, V., Couvé, E., Rusniok, C., Trieu-Cuot, P., and Glaser, P. (2009). Atypical association of DDE transposition with conjugation specifies a new family of mobile elements. Mol. Microbiol. *71*, 948–959.

Brouwer, M.S.M., Warburton, P.J., Roberts, A.P., Mullany, P., and Allan, E. (2011). Genetic organisation, mobility and predicted functions of genes on integrated, mobile genetic elements in sequenced strains of *Clostridium difficile*. PLoS One *6*, e23014.

Chee, G.J., and Takami, H. (2011). Alternative splicing by participation of the group II intron ORF in extremely halotolerant and alkaliphilic *Oceanobacillus iheyensis*. Microbes Environ. *26*, 54–60.

Diaz-Torres, M.L., McNab, R., Spratt, D.A., Villedieu, A., Hunt, N., Wilson, M., and Mullany, P. (2003). Novel tetracycline resistance determinant from the oral metagenome. Antimicrob. Agents Chemother. *47*, 1430–1432.

Djordjevic, S.P., Cain, A.K., Evershed, N.J., Falconer, L., Levings, R.S., Lightfoot, D., and Hall, R.M. (2009). Emergence and evolution of multiply antibiotic-resistant *Salmonella enterica* serovar Paratyphi B D-tartrate-utilizing strains containing SGI1. Antimicrob. Agents Chemother. *53*, 2319–2326.

van, Essen-Zandbergen, A., Smith, H., Veldman, K., and Mevius, D. (2007). Occurrence and characteristics of class 1, 2, and 3 integrons in *Escherichia coli*, *Salmonella* and *Campylobacter* spp. in the Netherlands. J. Antimicrob. Chemother. *59*, 746–750.

Febrer, M., McLay, K., Caccamo, M., Twomey, K.B., and Ryan, R.P. (2011). Advances in bacterial transcriptome and transposon insertion-site profiling using second-generation sequencing. Trends Biotechnol. *29*, 586-594.

Franciosa, G., Scalfaro, C., Di Bonito, P., Vitale, M., and Aureli, P. (2011). Identification of novel linear megaplasmids carrying a ß-lactamase gene in neurotoxigenic *Clostridium butyricum* Type E strains. PLoS One *6*, e21706.

Frost, L.S., and Koraimann, G. (2010). Regulation of bacterial conjugation: balancing opportunity with adversity. Future Microbiol. *5,* 1057–1071.

Grindley, N.D.F., Whiteson, K.L., and Rice, P.A. (2006). Mechanisms of site-specific recombination. Annu. Rev. Biochem. *75,* 567–605.

Haniford, D.B. (2006). Transpososome dynamics and regulation in Tn10 transposition. Crit. Rev. Biochem. Mol. Biol. *X,* 407–424.

Harshey, R.M., and Jayaram, M. (2007). The mu transpososome through a topological lens. Crit. Rev. Biochem. Mol. Biol. *41,* 387–405.

Jones, B.V., and Marchesi, J.R. (2007). Transposon-aided capture (TRACA) of plasmids resident in the human gut mobile metagenome. Nat. Methods *4,* 55–61.

Jones, B.V., Sun, F., and Marchesi, J.R. (2010). Comparative metagenomic analysis of plasmid encoded functions in the human gut microbiome. BMC Genomics *11,* 46.

Kurokawa, K., Itoh, T., Kuwahara, T., Oshima, K., Toh, H., Toyoda, A., Takami, H., Morita, H., Sharma, V.K., Srivastava, T.P., *et al.* (2007). Comparative metagenomics revealed commonly enriched gene sets in human gut microbiomes. DNA Res. *14,* 169–181.

Lévesque, C., Piché, L., Larose, C., and Roy, P.H. (1995). PCR mapping of integrons reveals several novel combinations of resistance genes. Antimicrob. Agents Chemother. *39,* 185–191.

Makarova, K.S., Haft, D.H., Barrangou, R., Brouns, S.J., Charpentier, E., Horvath, P., Moineau, S., Mojica, F.J., Wolf, Y.I., Yakunin, A.F., *et al.* (2011). Evolution and classification of the CRISPR-Cas systems. Nat. Rev. Microbiol. *9,* 467–477.

Martinez-Freijo, P., Fluit, A.C., Schmitz, F.J., Grek, V.S., Verhoef, J., and Jones, M.E. (1998). Class 1 integrons in Gram-negative isolates from different European hospitals and association with decreased susceptibility to multiple antibiotic compunds. J. Antimicrob. Chemother. *42,* 689–696.

Otto, M. (2010). *Staphylococcus aureus* toxin gene hitchhikes on a transferable antibiotic resistance element. Virulence *1,* 49–51.

Qu, A., Brulc, J.M., Wilson, M.K., Law, B.F., Theoret, J.R., Joens, L.A., Konkel, M.E., Angly, F., Dinsdale, E.A., Edwards, R.A., *et al.* (2008). Comparative metagenomics reveals host specific metavirulomes and horizontal gene transfer elements in the chicken cecum microbiome. PLoS One *3,* e2945.

Ravin, N.V. (2003). Mechanisms of replication and telomere resolution of the linear plasmid prophage N15. FEMS Microbiol. Lett. *221,* 1–6.

Roberts, A.P., and Mullany, P. (2009). A modular master on the move: the Tn916 family of mobile genetic elements. Trends Microbiol. *17,* 251–258.

Roberts, A.P., and Mullany, P. (2011). Tn916-like genetic elements: a diverse group of modular mobile elements conferring antibiotic resistance. FEMS Microbiol. Rev. *35,* 856-871.

Seville, L.A., Patterson, A.J., Scott, K.P., Mullany, P., Quail, M.A., Parkhill, J., Ready, D., Wilson, M., Spratt, D., and Roberts, A.P. (2009). Distribution of tetracycline and erythromycin resistance genes among human oral and fecal metagenomic DNA. Microb. Drug Resist. *15,* 159–166.

Scotta, *et al.* (2007). Exploring Integron Diversity. Poster 026 presented at: 9th Symposium on Bacterial Genetics and Ecology (BAGECO 9), p. 75.

Siguier, P., Filée, J., and Chandler, M. (2006). Insertion sequences in prokaryotic genomes. Curr. Opin. Microbiol. *9,* 526–531.

Smillie, C., Garcillán-Barcia, M.P., Francia, M.V., Rocha, E.P., de la Cruz, F. (2010). Mobility of plasmids. Microbiol. Mol. Biol. Rev. *74,* 434–452.

Smith, M.C., and Thorpe, H.M. (2002). Diversity in the serine recombinases. Mol. Microbiol. *44,* 299–307.

Stokes, H.W., Holmes, A.J., Nield, B.S., Holley, M.P., Nevalainen, K.M., Mabbutt, B.C., and Gillings. M.R. (2001). Gene cassette PCR: sequence-independent recovery of entire genes from environmental DNA. Appl. Environ. Microbiol. *67,* 5240–5246.

Suenaga, H., Koyama, Y., Miyakoshi, M., Miyazaki, R., Yano, H., Sota, M., Ohtsubo, Y., Tsuda, M., and Miyazaki, K. (2009). Novel organization of aromatic degradation pathway genes in a microbial community as revealed by metagenomic analysis. ISME J. *3,* 1335–1348.

Toro, N., Jiménez-Zurdo, J.I., and García-Rodríguez, F.M. (2007). Bacterial group II introns: not just splicing. FEMS Microbiol. Rev. *31,* 342–358.

de Vries, L.E., Vallès, Y., Agersø, Y., Vaishampayan, P.A., García-Montaner, A., Kuehl, J.V., Christensen, H., Barlow, M., and Francino, M.P. (2011). The gut as reservoir of antibiotic resistance: microbial diversity of tetracycline resistance in mother and infant. PLoS ONE *6,* e21644.

Wagner, A. (2006). Cooperation is fleeting in the world of transposable elements. PLoS Comput. Biol. *2,* e162.

Warburton, P., Allan, E., Hunter, S., Ward, J., Booth, V., Wade, W.G., and Mullany, P. (2011). Isolation of bacterial extra-chromosomal DNA from human dental plaque associated with periodontal disease, using transposon-aided capture (TRACA). FEMS Microbiol. Ecol. *78*, 349-354.

Wozniak, R.A., and Waldor, M.K. (2010). Integrative and conjugative elements: mosaic mobile genetic elements enabling dynamic lateral gene flow. Nat Rev Microbiol. *8*, 552–563.

Yao, J., and Lambowitz, A.M. (2007). Gene targeting in Gram-negative bacteria by use of a mobile group II intron ('Targetron') expressed from a broad-host-range vector. Appl. Environ. Microbiol. *73*, 2735–2743.

Horizontal Gene Transfer and Recombination in the Evolution of Antibiotic Resistance Genes

9

Miriam Barlow, Jared Caywood, Serena Lai, Joshua Finley and Chad Swanlund

Abstract

Recombination is a mechanism that leads to variation in antibiotic resistance genes. We review its importance in the recently emerged nosocomial pathogen *Acinetobacter baumannii*, where horizontal gene transfer (HGT) and recombination are major sources of variation for resistance phenotypes. We also present our results focusing on the plasmid-borne genes of bla_{CTX-M}, *ampC* and *qac*. These results show that random point mutations are not the only major source of variants among the plasmid-borne genes, but that homologous recombination is a major contributor of gene variants that can likewise add and remove resistance phenotypes from bacterial populations. Further studies of HGT and recombination are therefore important to determine how their effects can be manipulated to control the occurrence of resistance.

HGT is recognized as one of the most troublesome aspects of antibiotic resistance because it rapidly moves resistance genes among pathogenic, commensal, and environmental bacteria. Conjugation is particularly effective at spreading resistance genes broadly among bacteria. Within the Gram-negative bacteria, plasmids have disseminated genes encoding β-lactamases broadly throughout Enterobacteriaceae, and as far as *Neisseria* and *Haemophilus* (Saunders *et al.*, 1986). The timeframe for worldwide dissemination of a newly emerged resistance gene is typically less than 3 years from the initial deployment of an antibiotic (Maple *et al.*, 1990; Clancy *et al.*, 1997; Medeiros, 1997).

It makes sense that HGT effectively disseminates antimicrobial resistance among bacterial populations, because most commercially developed antimicrobials are derived from those naturally produced by microbes (Martin, 1998, 2000; Martin *et al.*, 2004). There was likely selection for antibiotic resistance in the pre-antibiotic era. For the antibiotics produced by microbes, resistance mechanisms already existed, long before the antibiotics were discovered or deployed. While bacteria do not carry innate resistance determinants for all antibiotics, they are able to rapidly acquire new resistance determinants from other bacteria through HGT. This has enabled bacteria to accumulate numerous resistance genes on plasmids or in resistance islands and become simultaneously resistant to multiple antibiotics.

Efforts to prevent the spread of antibiotic resistance through planned or restricted use of antibiotics have yielded mixed results. There have been numerous calls for conservative prescription of antibiotics to prolong the useful lifetime of antimicrobial agents (Livermore, 2005). To some extent, those calls have been met. In the UK, community antimicrobial prescribing was reduced by as much as 22% from 1995 to 2000 with a > 50% reduction in paediatric prescribing during the same period (Wise, 2004). In the US, similar reductions in some antimicrobial prescriptions have been observed (McCaig *et al.*, 2003). However, there is not a clear relationship between the reduced usage of antibiotics and the subsequent prevalence of resistance to them. In the UK there has been a steady decline in the frequency of penicillin resistant *Streptococcus pneumoniae* since usage of that antibiotic was reduced. However, macrolide resistance has remained steady in *S. pneumoniae* isolates since 1997 (~13%) (Livermore, 2004; Livermore, 2005) even though reduction of macrolide usage has also occurred. A particularly alarming trend has been the steady rise in β-lactam resistance among *Escherichia coli* in the UK regardless of reductions in β-lactam consumption (Livermore, 2005). This may result from unknown selection for resistance, the absence of fitness costs for carrying resistance, and/or over-representation of resistance genes through HGT.

Besides simply reducing the consumption of antibiotics, there have been several attempts to reduce the occurrence of resistance by cycling antibiotics, or rotating the order in which they are used (Gerding, 2000; Moss *et al.*, 2002; Warren *et al.*, 2004; Brown and Nathwani, 2005; Martinez *et al.*, 2006). In these studies, physicians in a hospital are only allowed to prescribe one type of antibiotic (fluoroquinolones, macrolides, β-lactams, aminoglycosides, etc.) if the health of the patient is not knowingly compromised, for a predetermined period of time. When the period for using one type of antibiotic has expired, all of the doctors begin prescribing the next. These efforts however have been designed empirically. This means that the relative fitness effects of resistance genes were not taken into consideration. Not surprisingly, these empiric cycling studies yielded varied results. In some cases resistance increased, and in others it decreased. Mathematical modelling studies for this type of cycling approach show that overall, empiric cycling schemes will either have no effect on resistance, or that they will actually increase the occurrence of resistance (Bergstrom *et al.*, 2004).

Orchestrating the prescribing practices of numerous doctors is a formidable and difficult effort, but has been achievable. The compliance of doctors to adhere to the rigid controls of cycling strategies shows that the medical community is ready and willing to make the changes necessary to reduce the occurrence of resistance. However, the failure of these efforts to reduce antibiotic resistance shows that the basic research underlying efforts to reduce resistance is lagging far behind the needs and preparation of the medical community. More basic research on the topic of antibiotic resistance is necessary to determine the best strategies.

While there are clearly no simple solutions to the antibiotic resistance problem, the general assumption that bacteria have infinite potential to acquire and express resistance genes is probably false. Although the plasticity of bacterial chromosomes allows bacteria to acquire and express numerous resistance genes via HGT, there are clear indications that limitations to the evolvability of bacteria do in fact exist. We do not yet understand those limitations with enough detail to develop from them more effective therapeutic practices. However, more knowledge about the population dynamics, genetic requirements, and trade-offs that regulate the overall impact of HGT on bacterial populations will likely yield the parameters critical for improved therapies.

As HGT and antibiotic resistance has been extensively reviewed multiple times (Davies and Davies, 2010; Palmer *et al.*, 2010; Smillie *et al.*, 2010), this chapter seeks to highlight important instances of HGT that have not yet been reviewed and to stress the role of genetic recombination as a molecular phenomenon contributing to the variation of resistance phenotypes in bacterial populations.

HGT in *Acinetobacter baumannii*

The pathogen *Acinetobacter baumannii* has emerged during the last decade as a nosocomial and opportunistic pathogen. It is currently responsible for 2–10% of nosocomial infections (Hanberger *et al.*, 1999; Jones *et al.*, 2004; Fournier *et al.*, 2006). It is particularly difficult to treat because it has many innate resistance mechanisms and readily acquires new resistance genes. There is overwhelming evidence that HGT is critical to the evolution of resistance in *A. baumannii* (Bou *et al.*, 2000; Peleg *et al.*, 2008; Ray *et al.*, 2009; Hornsey *et al.*, 2011; Smith *et al.*, 2011).

Over 17% of the ORFs in *A. baumannii* are located in 28 islands of foreign DNA. Sixteen of the islands are involved in pathogenesis while one appears to mainly harbour resistance genes (Smith *et al.*, 2011). Those genes encode resistance to β-lactams, aminoglycosides, fluoroquinolones, tetracyclines, chloramphenicol, rifampin, sulfonamides, heavy metals and ammonium (Peleg *et al.*, 2008; Smith *et al.*, 2011). Resistance genes are also distributed throughout the genome outside of any resistance island. The mechanism through which the foreign DNA contained in the islands enters *A. baumannii* is not known. It is possible that it enters on a plasmid, stably incorporates into the genome, and the plasmid is subsequently shed by the bacteria. It is also possible that *A. baumanii* is naturally transformable since its near relative *Acinetobacter baylii* is a highly transformable strain (Vaneechoutte *et al.*, 2006; Ray *et al.*, 2009).

A striking example of the rapid evolvability of *A. baumanii* is given in a study by Hornsey *et al.* (2011). They sequenced the genomes of two isolates of *A. baumanii* collected from the same patient immediately preceding and following a 1-week course of tigecycline therapy. They found that during the week-long course of treatment, resistance to tigecycline increased from a Minimum Inhibitory Concentration (MIC) of 0.5 μg/ml to 16 μg/ml. Genome sequencing of pre- and post-treatment strains revealed the acquisition of two plasmids, the occurrence of 18 SNPs and three deletions ranging from 15 to 44 kb in the latter isolate relative to the former (Hornsey *et al.*, 2011). While the effects of all the genetic changes have not been determined, one SNP probably resulted in the up-regulation of efflux, which led to tigecycline resistance.

Although tigecycline resistance evolved rapidly, aminoglycoside resistance diminished, probably as a result of one of the deletions. For tobramycin, the MIC decreased from > 32 μg/ml to 2 μg/ml; for amikacin from > 64 μg to 4 μg/ml; for gentamicin from > 32 μg/ml to 8 μg/ml. A direct cause–effect relationship between the occurrence of tigecycline resistance and the loss of aminoglycoside resistance seems unlikely, but they may indicate limitations to the overall resistance potential of *A. baumanii* (Hornsey *et al.*, 2011). It is important to understand the dynamics that exist between the particular mechanisms of HGT and selection to determine what limitations may exist in the evolutionary potential of *A. baumanii*. By understanding those limitations, it may be possible to inhibit the evolution of resistance among those bacteria.

Increasing occurrence of plasmids and recombination in plasmid-borne genes

The hefty fitness benefits attributable to resistance genes have probably increased the occurrence of HGT in bacteria, especially conjugation. In a surveillance study of pre-antibiotic era bacteria, a mere 19% of bacterial strains contained plasmids (Datta and Hughes, 1983).

In surveillance studies of more contemporary strains, anywhere from 58% to 100% carry plasmids (Souza *et al.*, 1999; Wan *et al.*, 2003) depending on the species surveyed. Although the collection methods and resistance detection assays varied in the studies used for this comparison (which may have introduced biases towards an increasing frequency of plasmids), few doubt that plasmid carriage is much more common among bacterial strains in the antimicrobial drug era (Mirkin *et al.*, 2003; Rainey and Cooper, 2004).

Many of the isolates examined in contemporary studies simultaneously carry multiple plasmids in bacterial strains (Souza *et al.*, 1999; Wan *et al.*, 2003; Barlow *et al.*, 2009). The abundance of plasmids and of some resistance genes upon plasmids has created a situation in which homologous recombination can occur among extra-chromosomal elements regularly. There is growing evidence that homologous recombination among plasmidic resistance alleles is an important source of variation in resistance genes that experience frequent horizontal transfer on plasmids.

Much of this evidence comes from plasmid-borne genes encoding β-lactamases. There are four classes of β-lactamases (A, B, C and D) that have been identified based on homology (Ambler, 1980). Although many genes from each class are found on chromosomes, those that constitute the greatest clinical threat are usually found on plasmids. The Class A β-lactamases have become the most frequently identified β-lactamases among antibiotic resistant isolates of bacteria (Decre *et al.*, 2004; Kim *et al.*, 2004; Kruger *et al.*, 2004; Lavigne *et al.*, 2004; Baraniak *et al.*, 2005; Brinas *et al.*, 2005; Ho *et al.*, 2005a,b; Schlesinger *et al.*, 2005; Tasli and Bahar, 2005). Likely reasons for the prevalence of Class A β-lactamases among bacteria is that they are expressed in both Gram-negative and Gram-positive bacteria, they have demonstrated exceptional evolutionary plasticity, and collectively, they confer resistance to all of the β-lactams that have been clinically introduced. The occurrence of Class A β-lactamases among human pathogens has seriously diminished the reliability of β-lactams.

TEMs

Generally speaking, the TEM β-lactamases have been the most commonly encountered Class A β-lactamases (Decre *et al.*, 2004; Kim *et al.*, 2004; Kruger *et al.*, 2004; Lavigne *et al.*, 2004; Baraniak *et al.*, 2005; Brinas *et al.*, 2005; Ho *et al.*, 2005a,b; Schlesinger *et al.*, 2005; Tasli and Bahar, 2005). The gene encoding TEM-1 ($bla_{\text{TEM-1}}$) was first identified in 1963 (Datta and Kontomichalou, 1965) and it appears to have spread across the globe in only a few years, although it is impossible to be certain about the exact timing of its spread because there were insufficient clinical isolates gathered between the 1940s and 1960s to determine the distribution of resistance genes in microbial populations. However, by the late 1970s and early 1980s, $bla_{\text{TEM-1}}$ was clearly the most prevalent plasmid-borne resistance gene in clinical microbial populations throughout the world (Matthew, 1979; Medeiros, 1984).

Most genes that encode TEMs (bla_{TEM}s) are separated by fewer than five mutations, and all of them are descended directly from $bla_{\text{TEM-1}}$. Most of the mutations that have occurred in the bla_{TEM}s have helped bacteria adapt to selective pressures imposed by antibiotics. For example, a substitution from arginine to serine at site 164 of the TEM-1 protein causes the mutant enzyme (TEM-12) to confer elevated levels of resistance to the β-lactam ceftazidime. An additional mutation of glutamic acid to lysine at site 240 in TEM-12 gives rise to a new allele (TEM-10) and further increases ceftazidime resistance (ceftazidime

MIC = 128 μg/ml) (Blazquez *et al.*, 1998). The genes that encode TEM-1, TEM-10, and TEM-12 are among the most commonly encountered resistance determinants in America (Jacoby and Bush, 2005).

The minute amount of sequence variation present in the *bla*$_{TEM}$ alleles made it difficult to investigate the occurrence of recombination among them. To do this, a new approach which only investigated silent variation was developed. The basic approach used was to make a phylogeny and infer the number of times each silent substitution independently arose, and then determine how likely it was that a mutation could independently arise that number of times given the overall mutation rate in the data set. The probability of a certain number of spontaneous silent point mutations occurring at a specific silent site was calculated using the Poisson distribution. A G-test was used to determine if there was a significant difference between the observed and expected numbers of silent mutations. That analysis showed that the occurrence of silent substitutions in *bla*$_{TEM}$s is greater than expected by spontaneous point mutation (Barlow *et al.*, 2009).

SHVs

A similar analysis was also performed for the *bla*$_{SHV}$ alleles in the same study (Barlow *et al.*, 2009). The SHVs are another group of class A β-lactamases that have been frequently observed in surveillance studies. As with the TEMs, numerous alleles encode unique SHV enzymes ($\approx$ 105), and the SHVs are all descended from a single ancestor (Hall and Barlow, 2004). The first *bla*$_{SHV}$ allele was detected in 1974 (Roupas and Pitton, 1974; Medeiros, 1997). Unlike *bla*$_{TEM}$s, the *bla*$_{SHV}$ s are present in the chromosome of nearly all *Klebsiella pneumoniae* isolates belonging to the KP1 group. Evidence suggests that *bla*$_{SHV}$ s have been chromosomally located since the pre–antimicrobial drug era (Haeggman *et al.*, 2004), and they may have been mobilized to plasmids up to four times, although the sequence divergence among them is insufficient to clearly resolve the independent mobilizations of the *bla*$_{SHV}$s.

CTX-Ms

We wanted to test the general importance of recombination as a mechanism of introduction of variation into other plasmid borne resistance genes. So we implemented the same approach to look for recombination in the *bla*$_{CTX\text{-}M}$s, and AmpC β-lactamases and qac genes.

The CTX-M group of enzymes is a globally distributed family of Class A extended spectrum β-lactamases (ESBLs) usually found on plasmids. They are of clinical importance because they exhibit various levels of activity against β-lactam antibiotics such as ceftazidime and cefotaxime (Bonnet, 2004). They are also important because the *bla*$_{CTX\text{-}M}$ family of genes has become broadly disseminated among Gram-negative bacteria such as *Salmonella enterica, Escherichia coli* and *Klebsiella pneumoniae* (Bou *et al.*, 2002; Ma *et al.*, 2002). In some locations, they are found more frequently than either the *bla*$_{TEM}$ or *bla*$_{SHV}$ genes (Morosini *et al.*, 2006; Empel *et al.*, 2008). *bla*$_{CTX\text{-}M}$ variants share > 90% identity and can be divided into five distinct groups (CTX-M groups 1, 2, 8, 9, and 25)(Bonnet, 2004). Within each group the sequence identity is > 94% (Bonnet, 2004). The CTX-Ms have functionally diverged as shown by various MICs of β-lactam antibiotics (Bonnet, 2004; Novais *et al.*, 2010). For example, expression of *bla*$_{CTX\text{-}M\text{-}15}$ confers greater ceftazidime resistance upon *E. coli* (MIC = 256 μg/ml), than expression of *bla*$_{CTX\text{-}M\text{-}3}$ (MIC = 32 μg/ml) (Bonnet, 2004) though they both belong to the CTX-M-1 group. Interestingly, the more distant *bla*$_{CTX\text{-}M\text{-}19}$ (part of the CTX-M-9 group) confers the next most resistance to ceftazidime (MIC = 128 μg/ml) suggesting either the convergent evolution of similar resistance phenotypes or horizontal exchange of the relevant nucleotides. While many of the substitutions that contribute to the phenotypic divergence of CTX-Ms are known, they may not have arisen from spontaneous point mutations, but from recombination.

We examined the independent occurrence of mutations in the *bla*$_{CTX\text{-}M}$s. It was necessary to aggregate the data from multiple groups to obtain a statistically significant sampling of mutations. After doing this, we found a significant difference between the distribution of independently occurring silent substitutions at specific sites and the overall occurrence of spontaneous point mutations for the analysis of the combined *bla*$_{CTX\text{-}M\text{-}2,8,25}$ groups, and the *bla*$_{CTX\text{-}M\text{-}9}$ group (Table 9.1). As with the other Class A β-lactamase genes, this implicates recombination as a likely source of genetic variation.

AmpCs

We looked outside the Class A β-lactamases by studying some members of the class C family (Ambler, 1980), namely the *ampC* antibiotic resistance gene family. This family is distantly related to class A, and homology is detectable only at the structural level, but it also confers resistance to ß-lactam antibiotics. The *ampC*s primarily confer resistance to cephalosporins. *ampC* is located in the chromosomes of the Enterobacteriaceae group, which includes *E. coli* and its close relatives. In the late 1980s an *ampC* gene was first detected on plasmids in resistant strains of bacteria (Bauernfeind *et al.*, 1989), and in 1990 an *ampC* gene from *Citrobacter freundii* was first observed on a plasmid (Bauernfeind *et al.*, 1990). That allele, *bla*$_{CMY\text{-}2}$, has since become distributed throughout several species of bacteria worldwide and is able to confer resistance to some antibiotics to which the TEM alleles cannot. Although a handful of alleles that are descended from *bla*$_{CMY\text{-}2}$ and that differ in amino acid sequence have been isolated, they do not appear to have evolved any new phenotypes (Barlow and Hall, 2002). While it is possible that there has not been sufficient time and/or selective pressure for *bla*$_{CMY\text{-}2}$ to give rise to dramatically different alleles, it is also possible that *bla*$_{CMY\text{-}2}$ lacks the evolutionary plasticity that has been observed in the TEM ß-lactamases.

We examined the independent substitutions in the *C. freundii* plasmid-borne *ampC* alleles. As for the Class A β-lactamases, the distribution of independent substitutions in the *ampC* genes was significantly different than that expected for spontaneous point mutations (Table 9.2) which indicates that recombination is a likely source of variation among many plasmid-borne β-lactamase alleles.

QACs

We extended the study farther to include a completely unrelated group of plasmid-borne resistance genes. The *qac* gene family shares no homology with any β-lactamases and can therefore demonstrate whether recombination has general importance among resistance genes. The *qac* genes encode for efflux pumps that lead to resistance to quaternary ammonium compounds (QAC). In the 1930s QACs were first used as biocides and have become a major part of common household cleaning agents. Since the implementation of these biocides, a number of plasmid-borne genes have been found that encode resistance to QACs. We analysed the patterns of silent substitutions in the *qac* genes. We also found a significant difference between the distribution of independent substitutions and that expected for spontaneous point mutations in the analysis of the *qac* groups (Table 9.3). There were

Table 9.1 Analysis of bla_{CTX-M} silent sites

	CTX-M 1 (i)[a]			CTX-M 2, 8, 25 (ii)[b]			CTX-M 9 (iii)[c]		
Number of mutations	**Poisson probability**	**Expected number of sites**	**Observed number of sites**	**Poisson probability**	**Expected number of sites**	**Observed number of sites**	**Poisson probability**	**Expected number of sites**	**Observed number of sites**
0	0.919	270.033	269	0.249	54.303	44	0.829	252.025	264
1	0.078	22.962	25	0.346	75.477	84	0.155	47.255	28
2	–	–	–	0.241	52.453	59	0.015	4.430	8
3	–	–	–	0.111	24.302	23	0.001	0.277	3
4	–	–	–	0.039	8.444	8	$4.3 \times 10{-}5$	0.013	1

[a] χ^2 *P*-value = 0.1390.
[b] χ^2 *P*-value = 0.041.
[c] χ^2 *P*-value = 1.47×10^{-5}.

Table 9.2 Analysis of *ampC* silent sites

Number of mutations	Poisson probability[a]	Expected number of sites[a]	Observed number of sites[a]
0	0.782	216.633	284
1	0.192	53.250	60
2	0.024	6.544	11
3	0.002	0.536	2

[a] *P*-value = 2.35453×10^{-34}.

Table 9.3 Analysis of *qac* Silent Sites

Number of mutations	Poisson probability	Expected number of sites	Observed number of sites[a]
0	0.00307	0.085	16
1	0.0178	0.497	0
2	0.0514	1.439	4
3	0.0991	2.775	0
4	0.143	4.015	0
5	0.166	4.646	1
6	0.160	4.480	0
7	0.132	3.703	0
8	0.0956	2.678	0
9	0.0615	1.721	0
10	0.0356	0.996	0
11	0.0187	0.524	1
12	0.00902	0.253	0
13	0.00401	0.112	0
14	0.00166	0.046	0
15	0.000640	0.018	0
16	0.000231	0.006	0
17	7.88×10^{-5}	0.002	0
18	2.53×10^{-5}	0.001	1
19	7.71×10^{-6}	2.158×10^{-4}	0
20	2.23×10^{-6}	6.24×10^{-5}	0
21	6.14×10^{-7}	1.72×10^{-5}	1
22	1.62×10^{-7}	4.52×10^{-6}	0
23	4.06×10^{-8}	1.14×10^{-6}	0
24	9.80×10^{-9}	2.74×10^{-7}	0
25	2.27×10^{-9}	6.35×10^{-8}	0
26	5.05×10^{-10}	1.41×10^{-8}	0
27	1.08×10^{-10}	3.03×10^{-9}	0
28	2.23×10^{-11}	6.26×10^{-10}	0
29	4.46×10^{-12}	1.25×10^{-10}	0
30	8.60×10^{-13}	2.41×10^{-11}	1
31	1.60×10^{-13}	4.49×10^{-12}	0
32	2.90×10^{-14}	8.12×10^{-13}	0
33	5.09×10^{-15}	1.42×10^{-13}	1
34	8.66×10^{-16}	2.42×10^{-14}	0
35	1.43×10^{-16}	4.01×10^{-15}	0
36	2.30×10^{-17}	6.44×10^{-16}	0
37	3.60×10^{-18}	1.01×10^{-16}	0
38	5.47×10^{-19}	1.53×10^{-17}	0
39	8.12×10^{-20}	2.27×10^{-18}	1

[a] χ^2 *P*-value $= 3.45496 \times 10^{-79}$.

some substitutions that had occurred numerous (up to 39) times, which was more than we observed for any of the other genes. As there were not particularly more *qac* alleles than other alleles included in the analysis, the most likely reason is probably that those alleles have existed on plasmids longer than the other alleles and have therefore had more opportunities to experience recombination.

Qnr

Overall, the results of this study illustrate that homologous recombination is a major source of variation for antimicrobial resistance genes. Additionally, a study by Baquirin and Barlow suggests that events such as genetic drift and recombination may influence the divergence of *qnr* (plasmidic fluoroquinolone resistance) genes to a greater extent than directional selection (Baquirin and Barlow, 2008). The widespread occurrence of recombination in different gene types illustrates the importance of this evolutionary mechanism for bacteria.

Conclusions

While the generation of variation through homologous recombination can create new resistance phenotypes, it can also remove resistance phenotypes from bacterial populations. Analysis of the co-occurrence of Class A β-lactamases in plasmids shows that multiple alleles of the same gene co-occur less often than expected in bacteria (Barlow *et al.*, 2009). This result was also supported by an experimental study in which homologous recombination removed resistance alleles that shared nearly identical DNA sequence but conferred very different phenotypes (Mroczkowska and Barlow, 2008). By understanding the occurrence of homologous recombination of plasmids it may be possible to predict what dynamics will lead to the generation of new resistance phenotypes and when recombination will eliminate resistance phenotypes from bacterial populations.

It is clear from these examples that HGT and homologous recombination of plasmids are important factors in the evolution of resistance. However, it is not clear that their role is always to promote the evolution or the dissemination of new resistance phenotypes. Through further studies of HGT and recombination and their relation to antibiotic resistance, it may be possible to identify strategies for preventing the occurrence of antibiotic resistance.

Acknowledgement

NIH grant 1R15GM090164-01A1 funded this research.

References

Ambler, R.P. (1980). The structure of beta-lactamases. Philos. Trans. R. Soc. Lond. B Biol. Sci. *289*, 321–331.

Baquirin, M.H., and Barlow, M. (2008). Evolution and recombination of the plasmidic qnr alleles. J. Mol. Evol. *67*, 103–110.

Baraniak, A., Fiett, J., Mrowka, A., Walory, J., Hryniewicz, W., and Gniadkowski, M. (2005). Evolution of TEM-type extended-spectrum beta-lactamases in clinical Enterobacteriaceae strains in Poland. Antimicrob. Agents Chemother. *49*, 1872–1880.

Barlow, M., and Hall, B.G. (2002). Origin and evolution of the AmpC β-lactamases of *Citrobacter freundii*. Antimicrob. Agents Chemother. *46*, 1190–1198.

Barlow, M., Fatollahi, J., and Salverda, M. (2009). Evidence for recombination among the alleles encoding TEM and SHV beta-lactamases. J. Antimicrob. Chemother. *63*, 256–259.

Bauernfeind, A., Chong, Y., and Schweighart, S. (1989). Extended broad spectrum beta-lactamase in *Klebsiella pneumoniae* including resistance to cephamycins. Infection *17*, 316–321.

Bauernfeind, A., Grimm, H., and Schweighart, S. (1990). A new plasmidic cefotaximase in a clinical isolate of *Escherichia coli*. Infection *18*, 294–298.

Bergstrom, C.T., Lo, M., and Lipsitch, M. (2004). Ecological theory suggests that antimicrobial cycling will not reduce antimicrobial resistance in hospitals. Proc. Natl. Acad. Sci. U.S.A. *101*, 13285–13290.

Blazquez, J., Negri, M.C., Morosini, M.I., Gomez-Gomez, J.M., and Baquero, F. (1998). A237T as a modulating mutation in naturally occurring extended-spectrum TEM-type beta-lactamases. Antimicrob. Agents Chemother. *42*, 1042–1044.

Bonnet, R. (2004). Growing group of extended-spectrum beta-lactamases: the CTX-M enzymes. Antimicrob. Agents Chemother. *48*, 1–14.

Bou, G., Oliver, A., and Martinez-Beltran, J. (2000). OXA-24, a novel class D beta-lactamase with carbapenemase activity in an *Acinetobacter baumannii* clinical strain. Antimicrob. Agents Chemother. *44*, 1556–1561.

Bou, G., Cartelle, M., Tomas, M., Canle, D., Molina, F., Moure, R., Eiros, J.M., and Guerrero, A. (2002). Identification and broad dissemination of the CTX-M-14 beta-lactamase in different *Escherichia coli* strains in the northwest area of Spain. J. Clin. Microbiol. *40*, 4030–4036.

Brinas, L., Lantero, M., de Diego, I., Alvarez, M., Zarazaga, M., and Torres, C. (2005). Mechanisms of resistance to expanded-spectrum cephalosporins in *Escherichia coli* isolates recovered in a Spanish hospital. J. Antimicrob. Chemother. *56*, 1107–1110.

Brown, E.M., and Nathwani, D. (2005). Antibiotic cycling or rotation: a systematic review of the evidence of efficacy. J. Antimicrob. Chemother. *55*, 6–9.

Clancy, J., Dib-Hajj, F., Petitpas, J.W., and Yuan, W. (1997). Cloning and characterization of a novel macrolide efflux gene, mreA, from *Streptococcus agalactiae*. Antimicrob. Agents Chemother. *41*, 2719–2723.

Datta, N., and Hughes, V.M. (1983). Plasmids of the same Inc groups in Enterobacteria before and after the medical use of antibiotics. Nature *306*, 616–617.

Datta, N., and Kontomichalou, P. (1965). Penicillinase synthesis controlled by infectious R factors in Enterobacteriacae. Nature *208*, 239–241.

Davies, J., and Davies, D. (2010). Origins and evolution of antibiotic resistance. Microbiol. Mol. Biol. Rev. *74*, 417–433.

Decre, D., Burghoffer, B., Gautier, V., Petit, J.C., and Arlet, G. (2004). Outbreak of multi-resistant *Klebsiella oxytoca* involving strains with extended-spectrum beta-lactamases and strains with extended-spectrum activity of the chromosomal beta-lactamase. J. Antimicrob. Chemother. *54*, 881–888.

Empel, J., Baraniak, A., Literacka, E., Mrowka, A., Fiett, J., Sadowy, E., Hryniewicz, W., and Gniadkowski, M. (2008). Molecular survey of beta-lactamases conferring resistance to newer beta-lactams in Enterobacteriaceae isolates from Polish hospitals. Antimicrob. Agents Chemother. *52*, 2449–2454.

Fournier, P.E., Vallenet, D., Barbe, V., Audic, S., Ogata, H., Poirel, L., Richet, H., Robert, C., Mangenot, S., Abergel, C., *et al.* (2006). Comparative genomics of multidrug resistance in *Acinetobacter baumannii*. PLoS Genet. *2*, e7.

Gerding, D.N. (2000). Antimicrobial cycling: lessons learned from the aminoglycoside experience. Infect. Control Hosp. Epidemiol. *21*, S12–S17.

Haeggman, S., Lofdahl, S., Paauw, A., Verhoef, J., and Brisse, S. (2004). Diversity and evolution of the class A chromosomal beta-lactamase gene in *Klebsiella pneumoniae*. Antimicrob. Agents Chemother. *48*, 2400–2408.

Hall, B.G., and Barlow, M. (2004). Evolution of the serine beta-lactamases: past, present and future. Drug Resist. Updat. *7*, 111–123.

Hanberger, H., Garcia-Rodriguez, J.A., Gobernado, M., Goossens, H., Nilsson, L.E., and Struelens, M.J. (1999). Antibiotic susceptibility among aerobic Gram-negative bacilli in intensive care units in 5 European countries. French and Portuguese ICU Study Groups. JAMA *281*, 67–71.

Ho, P.L., Ho, A.Y., Chow, K.H., Wong, R.C., Duan, R.S., Ho, W.L., Mak, G.C., Tsang, K.W., Yam, W.C., and Yuen, K.Y. (2005a). Occurrence and molecular analysis of extended-spectrum β-lactamase-producing *Proteus mirabilis* in Hong Kong, 1999–2002. J. Antimicrob. Chemother. *55*, 840–845.

Ho, P.L., Shek, R.H., Chow, K.H., Duan, R.S., Mak, G.C., Lai, E.L., Yam, W.C., Tsang, K.W., and Lai, W.M. (2005b). Detection and characterization of extended-spectrum beta-lactamases among bloodstream isolates of Enterobacter spp. in Hong Kong, 2000–2002. J. Antimicrob. Chemother. *55*, 326–332.

Hornsey, M., Loman, N., Wareham, D.W., Ellington, M.J., Pallen, M.J., Turton, J.F., Underwood, A., Gaulton, T., Thomas, C.P., Doumith, M., *et al.* (2011). Whole-genome comparison of two *Acinetobacter baumannii* isolates from a single patient, where resistance developed during tigecycline therapy. J. Antimicrob. Chemother. *66*, 1499–1503.

Jacoby, G.A., and Bush, K. (2005). B-lactam resistance in the 21st century. In Frontiers in Antimicrobial Resistance: a Tribute to Stuart, B Levy, White, D.G., Alekshun, M.N., and McDermott, P.F., eds (ASM Press, Washington, DC), pp. 53–65.

Jones, M.E., Draghi, D.C., Thornsberry, C., Karlowsky, J.A., Sahm, D.F., and Wenzel, R.P. (2004). Emerging resistance among bacterial pathogens in the intensive care unit – a European and North American Surveillance study (2000–2002). Ann. Clin. Microbiol. Antimicrob. *3*, 14.

Kim, S., Kim, J., Kang, Y., Park, Y., and Lee, B. (2004). Occurrence of extended-spectrum beta-lactamases in members of the genus *Shigella* in the Republic of Korea. J. Clin. Microbiol. *42*, 5264–5269.

Kruger, T., Szabo, D., Keddy, K.H., Deeley, K., Marsh, J.W., Hujer, A.M., Bonomo, R.A., and Paterson, D.L. (2004). Infections with nontyphoidal *Salmonella* species producing TEM-63 or a novel TEM enzyme, TEM-131, in South Africa. Antimicrob. Agents Chemother. *48*, 4263–4270.

Lavigne, J.P., Bouziges, N., Chanal, C., Mahamat, A., Michaux-Charachon, S., and Sotto, A. (2004). Molecular epidemiology of Enterobacteriaceae isolates producing extended-spectrum beta-lactamases in a French hospital. J. Clin. Microbiol. *42*, 3805–3808.

Livermore, D. (2004). Can better prescribing turn the tide of resistance? Nat. Rev. Microbiol. *2*, 73–78.

Livermore, D.M. (2005). Minimising antibiotic resistance. Lancet Infect. Dis. *5*, 450–459.

Ma, L., Ishii, Y., Chang, F.Y., Yamaguchi, K., Ho, M., and Siu, L.K. (2002). CTX-M-14, a plasmid-mediated CTX-M type extended-spectrum beta-lactamase isolated from *Escherichia coli*. Antimicrob. Agents Chemother. *46*, 1985–1988.

McCaig, L.F., Besser, R.E., and Hughes, J.M. (2003). Antimicrobial drug prescription in ambulatory care settings, United States, 1992–2000. Emerg. Infect. Dis. *9*, 432–437.

Maple, P., Brumfitt, W., and Hamilton-Miller, J.M. (1990). A review of the antimicrobial activity of the fluoroquinolones. J. Chemother. *2*, 280–294.

Martin, J.F. (1998). New aspects of genes and enzymes for beta-lactam antibiotic biosynthesis. Appl. Microbiol. Biotechnol. *50*, 1–15.

Martin, J.F. (2000). Molecular control of expression of penicillin biosynthesis genes in fungi: regulatory proteins interact with a bidirectional promoter region. J. Bacteriol. *182*, 2355–2362.

Martin, J.F., Ullan, R.V., and Casqueiro, J. (2004). Novel genes involved in cephalosporin biosynthesis: the three-component isopenicillin N epimerase system. Adv. Biochem. Eng. Biotechnol. *88*, 91–109.

Martinez, J.A., Nicolas, J.M., Marco, F., Horcajada, J.P., Garcia-Segarra, G., Trilla, A., Codina, C., Torres, A., and Mensa, J. (2006). Comparison of antimicrobial cycling and mixing strategies in two medical intensive care units. Crit. Care Med. *34*, 329–336.

Matthew, M. (1979). Plasmid-mediated beta-lactamases of Gram-negative bacteria: properties and distribution. J. Antimicrob. Chemother. *5*, 349–358.

Medeiros, A.A. (1984). Beta-lactamases. Br. Med. Bull. *40*, 18–27.

Medeiros, A.A. (1997). Evolution and dissemination of beta-lactamases accelerated by generations of beta-lactam antibiotics. Clin. Infect. Dis. *24*, S19–45.

Mirkin, B.G., Fenner, T.I., Galperin, M.Y., and Koonin, E.V. (2003). Algorithms for computing parsimonious evolutionary scenarios for genome evolution, the last universal common ancestor and dominance of horizontal gene transfer in the evolution of prokaryotes. BMC Evol. Biol. 3, 2.

Morosini, M.I., Garcia-Castillo, M., Coque, T.M., Valverde, A., Novais, A., Loza, E., Baquero, F., and Canton, R. (2006). Antibiotic coresistance in extended-spectrum-beta-lactamase-producing Enterobacteriaceae and *in vitro* activity of tigecycline. Antimicrob. Agents Chemother. *50*, 2695–2699.

Moss, W.J., Beers, M.C., Johnson, E., Nichols, D.G., Perl, T.M., Dick, J.D., Veltri, M.A., and Willoughby, R.E., Jr. (2002). Pilot study of antibiotic cycling in a pediatric intensive care unit. Crit. Care Med. *30*, 1877–1882.

Mroczkowska, J.E., and Barlow, M. (2008). Recombination and selection can remove blaTEM alleles from bacterial populations. Antimicrob. Agents Chemother. *52*, 3408–3410.

Novais, A., Comas, I., Baquero, F., Canton, R., Coque, T.M., Moya, A., Gonzalez-Candelas, F., and Galan, J.C. (2010). Evolutionary trajectories of beta-lactamase CTX–M-1 cluster enzymes: predicting antibiotic resistance. PLoS Pathog. *6*, e1000735.

Palmer, K.L., Kos, V.N., and Gilmore, M.S. (2010). Horizontal gene transfer and the genomics of enterococcal antibiotic resistance. Curr. Opin. Microbiol. *13*, 632–639.

Peleg, A.Y., Seifert, H., and Paterson, D.L. (2008). *Acinetobacter baumannii*: emergence of a successful pathogen. Clin. Microbiol. Rev. *21*, 538–582.

Rainey, P.B., and Cooper, T.F. (2004). Evolution of bacterial diversity and the origins of modularity. Res. Microbiol. *155*, 370–375.

Ray, J.L., Harms, K., Wikmark, O.G., Starikova, I., Johnsen, P.J., and Nielsen, K.M. (2009). Sexual isolation in *Acinetobacter baylyi* is locus-specific and varies 10,000-fold over the genome. Genetics *182*, 1165–1181.

Roupas, A., and Pitton, J.S. (1974). R factor-mediated and chromosomal resistance to ampicillin in *Escherichia coli*. Antimicrob. Agents Chemother. *5*, 186–191.

Saunders, J.R., Hart, C.A., and Saunders, V.A. (1986). Plasmid-mediated resistance to beta-lactam antibiotics in Gram-negative bacteria: the role of in-vivo recyclization reactions in plasmid evolution. J. Antimicrob. Chemother. *18(Suppl C)*, 57–66.

Schlesinger, J., Navon-Venezia, S., Chmelnitsky, I., Hammer-Munz, O., Leavitt, A., Gold, H.S., Schwaber, M.J., and Carmeli, Y. (2005). Extended-spectrum beta-lactamases among *Enterobacter* isolates obtained in Tel Aviv, Israel. Antimicrob. Agents Chemother. *49*, 1150–1156.

Smillie, C., Garcillan-Barcia, M.P., Francia, M.V., Rocha, E.P., and de la Cruz, F. (2010). Mobility of plasmids. Microbiol. Mol. Biol. Rev. *74*, 434–452.

Smith, M.G., Gianoulis, T.A., Pukatzki, S., Mekalanos, J.J., Ornston, L.N., Gerstein, M., and Snyder, M. (2011). New insights into *Acinetobacter baumanii* pathogenesis reealed by high-density pyrosquencing and transposon mutagenesis. Genes Dev. *21*, 601–614.

Souza, V., Rocha, M., Valera, A., and Eguiarte, L.E. (1999). Genetic structure of natural populations of *Escherichia coli* in wild hosts on different continents. Appl. Environ. Microbiol. *65*, 3373–3385.

Tasli, H., and Bahar, I.H. (2005). Molecular characterization of TEM- and SHV-derived extended-spectrum beta-lactamases in hospital-based Enterobacteriaceae in Turkey. Jpn J. Infect. Dis. *58*, 162–167.

Vaneechoutte, M., Young, D.M., Ornston, L.N., De Baere, T., Nemec, A., Van Der Reijden, T., Carr, E., Tjernberg, I., and Dijkshoorn, L. (2006). Naturally transformable Acinetobacter sp. strain ADP1 belongs to the newly described species *Acinetobacter baylyi*. Appl. Environ. Microbiol. *72*, 932–936.

Wan, K.F., Radu, S., Cheah, Y.K., Benjamin, P.G., Ling, C.M., Hon, S.F., Harun, A., Lo, S.K., and Liew, P.K. (2003). Antibiotic resistance, plasmid profile and RAPD-PCR analysis of enteropathogenic *Escherichia coli* (EPEC) clinical isolates. Southeast Asian J. Trop. Med. Public Health *34*, 620–626.

Warren, D.K., Hill, H.A., Merz, L.R., Kollef, M.H., Hayden, M.K., Fraser, V.J., and Fridkin, S.K. (2004). Cycling empirical antimicrobial agents to prevent emergence of antimicrobial-resistant Gram-negative bacteria among intensive care unit patients. Crit. Care Med. *32*, 2450–2456.

Wise, R. (2004). The relentless rise of resistance? J. Antimicrob. Chemother. *54*, 306–310.

Index

A

C

D

F

H

I

N

Q

R

S

V

W

X

Y

Z

Current books of interest